George H Townsend, Felix J. Levy, Harry G. Nicks, George Clinton Crandall

The Relation of Food to Health and Premature Death

George H Townsend, Felix J. Levy, Harry G. Nicks, George Clinton Crandall

The Relation of Food to Health and Premature Death

ISBN/EAN: 9783744646024

Printed in Europe, USA, Canada, Australia, Japan

Cover: Foto ©berggeist007 / pixelio.de

More available books at **www.hansebooks.com**

THE

Relation of Food to Health
and Premature Death.

By GEO. H. TOWNSEND, LL.B.,

With the collaboration of

FELIX J. LEVY, A.M., M.D.

H. G. NICKS, M.D.,

Lecturer on Hygiene, Marion Sims College of Medicine; Attending Physician Woman's Hospital; Director Physical Department Y. M. C. A., St. Louis.

AND

GEO. CLINTON CRANDALL, B. S. M. D.

Professor of General Medicine, Marion Sims College of Medicine, St. Louis.

ST. LOUIS, MO.
WITT PUBLISHING COMPANY,
1897.

PREFACE.

This is an age of wonderful strides in production, but we fear that man, in improving everything else, has, in a great measure lost sight of himself. To the scientist who understands something of the wonderful development of nature, when free from hindrance, there is nothing so utterly astonishing as the weakness and folly of the human race. Believing that ignorance of self is the mother of our devouring evils—disease, vice and crime—the author, with the assistance of his collaborators, has undertaken to blaze out a road to a better and higher life, and however painstaking the effort, it would be too much to expect that our labors would produce results that approached the ideal. It is hoped, however, that this book will be of service in pointing out the devious windings into which appetite and surrounding influences often allure the thoughtless—resulting in their discomfiture and premature death.

Physical, mental and moral perfection can only exist when our lives come into harmony with natural laws, and when we cease to antagonize nature, the work will be done.

If we have made plain the most common transgressions of nature and how to minimize their effects, our purpose will have been accomplished.

GEO. H. TOWNSEND.

St. Louis, Mo., July, 1897.

CONTENTS.

CONTENTS.

CHAPTER I.

AFFLICTIONS AND PREMATURE DEATH RESULT FROM IGNORANCE.

Each age has its philanthropists, those who toil not merely for their own aggrandizement, but for the bettering, the uplifting of the human race. These make the world better for having lived in it. Such ought to be the desire of every person, and while it is sad to say that it is not the fact, this is truly an age in which proportionately more people are interested in the welfare of the race than at any period in the history of the world. It ought therefore to naturally follow, that the world should now be making greater strides towards ideal conditions than ever before. Perhaps we are doing this; but it is a matter which many well informed people would gravely question. No doubt but that all will agree, that no movement has ever been inaugurated for the elevation of man, which shows results commensurate with the effort expended. Why is this? There can be but one answer; it is because all efforts of every kind and character have been directed toward relieving, curing or reforming the individual; whereas, had all the efforts of even ten generations been directed toward preventing evil and disease we would now have an ideal race; but this would require a volume of itself, whereas the object sought, is to call attention to, and emphasize the fact that, it is ignorance of the laws governing our physical existence, creation, birth, and living, that makes reform movements necessary—movements which seek merely to overcome results of forces without dealing with their causes. This would be denied by nearly every man or woman engaged in trying to reform the world.

To illustrate; there are three ways of dealing with drunkenness:

  (1) Drugs or dipsomania cures.
  (2) Moral influences, signing the pledge, etc.
  (3) Prohibition—restraint by law.

Now the advocates of each of these methods claim that they deal with the causes of drunkenness, and yet men will not stay cured, nor keep the pledge, neither will they regard the law, and it is a lamentable fact that the army of drunkards is being constantly recruited from the families of the most zealous temperance advocates of the country. Something is wrong; for neither drugs, moral suasion nor law, have succeeded in arresting drunkeness, because the diet and habits of the people cause them to transmit nervous tendencies to each generation, and these are continually crying out for stimulation of some kind.

The truth of the matter is that most reformers have ignored the fact, *that the body, to a great extent, controls the mind, and therefore, the conduct. We are animals, without the governing instincts of brutes, and so limited in reason and knowledge, as to be unable to properly regulate our conduct.* A man born under proper conditions, and given correct knowledge of living, will need neither moral suasion nor prohibition to keep him from the liquor habit, and so far as this world is concerned, will not need any reform movement, or dread of future punishment to make him a good citizen.

Some years ago, the writer became profoundly interested in social and economic questions and the elevation of the race through popular edcuation, especially on the principles of living and the proper relation of the individual to society. After studying every phase of human conditions and character, he became imbued with the belief that more good could be accomplished by teaching the people the principles governing their physical existence than could be done in any other way. In order to bring this knowledge to them, all sources of information have been sought, and especially from those physicians whose training and experience warrant them in speaking with some degree of assurance. The facts obtained from all sources are given as one interview.

Venturing forth in quest of knowledge, the bookmaker sought the most learned specialists who have made a life study of food in its relation to health and disease. The first interview with a distinguished specialist in diseases of the stomach began with the explanation that the bookmaker was desirous of dispelling some of the darkness in which our physical existence is enveloped. "That," replied the doctor "is a great task, and worthy of the best effort that can be given it."

WHERE ALL OUR ILLS COME FROM.

"Some people declare that the masses cannot be moved to a more rational mode of living."

"That is worse than the facts warrant, for there are people who are really anxious to learn more about the principles which govern their existence."

"But isn't it also true that many people don't care to know anything?"

"Yes, and it is a strange thing that people are willing to suffer pain, lose the time of being sick, and then have to pay their money to doctors, when it could all be avoided."

"If that be true, what will become of the doctors?"

"Oh well, the people could better afford to pay the doctors to keep them well like the chinese than have to pay them and be sick; but when the people learn that their ills do not come from God, or from Adam, or even from nature, and learn that they are mostly self inflicted, or at farthest, come from their parents, they will learn how to dispense with both physic and physician."

"That's a good deal to say."

"Yes," said the doctor, "but I do not hesitate to say that it is not more than the truth."

"How could the people be brought to such a condition?

"They must first realize their individual responsibility."

"How can they be brought to that?"

"By presenting facts to them in a clear and forcible manner, which we will do. Now suppose I say, that practically every person commits suicide, and that a great many also commit manslaughter."

"If you did I should say that you were either jesting or crazy."

"But it is a fact, which I will prove. Let me ask you what would happen if you were to drive recklessly through the streets and in doing so run over several people and maim or kill them?"

"I would be arrested for manslaughter."

"Yes, and it would make no difference, except in the degree of punishment, whether you did it wilfully or negligently, you would be liable both civilly and criminally for injuring or killing another in such a manner. Now suppose your family had typhoid fever, and you should throw out some excrement and poison the well or stream from which your neighbor is supplied and sickness or death results, (which has occurred thousands of times) would it not be just as bad as to negligently kill him by an infectious disease, as to kill him by negligently running over him?"

"Yes, I suppose it would, only the proof more difficult."

"But that does not alter the fact, nor atone for the criminality of negligently spreading infectious diseases and death, which is continually being done, but this is not worse than other life destroying negligence which is even more appalling in effect."

"I can not deny your facts nor your conclusions, for they are overwhelming."

"Let me give you another illustration. A friend of mine was called to see a child four years old who had a serious intestinal disorder. The child was soon convalescent, and the doctor said his visits need not continue, but at the same time cautioned the parents to be exceedingly careful about the child's diet for 'two or three weeks.' "

"What happened?"

"Well, the day after the doctor's last visit the family had saurkrout for dinner and allowed the child to eat all it wanted under the belief that it would not hurt it."

"And that probably killed it."

"Yes, it was taken ill at once and the doctor called, but when he found what it had eaten, and the condition the child was in, he bluntly told its parents that they had killed their child."

"That was certainly a most distressing thing for the parents."

"So it was, but not worse than occurs in nearly every family, although it may not be quite so immediately apparent."

"I suppose they excuse themselves on the ground that they did not know any better?"

"Very likely, *but that is a poor excuse, for the knowledge could have been obtained.* This suggests the question· Is a person who is so careless and indifferent to things pertaining to life and health, that he.kills some one less culpable than one who negligently takes life in some other way?" .

"Doctor you put things so strongly, I think you could almost arouse the dead,and yet every word you have said is true."

"But what I have said only relates to the injury inflicted on others, and bad as it may seem, *self-destruction is far more common and its effects almost endless.*"

"You don't say! What are you trying to make me believe we are?"

"Oh, don't get excited, for I want to bring out another point by asking you a question."

"What is it?"

"How would you define suicide?"

"Well if one wilfully destroys his life, by making it shorter than nature intended, that would be suicide."

"It would make no difference whether the method was quick or slow, would it?"

"No, if it did it would be making a distinction without a difference."

"Now I suppose that everyone will admit that the moral law is higher than the law of the state, and if it recognizes negligence that injures another, the same as if wilfully done, the moral responsibility must be equally great. Here is another thing, it must be true, that each individual exists for a purpose, and if so who can measure the wrong of thwarting nature, by cutting off the natural term of life?"

"Doctor, you have proven that self-destruction is universal, and now you have gone farther and proven that it is practically suicide."

"Yes; wrongs are great or small in proportion to their effect, and it is difficult to see wherein an untimely death from one cause, that could have been avoided, is not as bad as from any other.

If the laws of our being were not so grossly violated one hundred years would be an average duration of life, and a hundred and fifty years not uncommon. The ordinary diseases of life should be wholly unknown, and though it may shock our slumbering senses, the facts make it necessary to say, that we take our own lives and are none the less culpable, because we do it ignorantly—the ignorance of negligence and careless indifference."

"That is good reasoning, and it is very strange that no one has ever written of it before."

"Yes it is, and the quotation from Shakespeare's Mid-Summer Night's Dream: 'what fools these mortals be' might be aptly applied. Just think, a young man will spend six or eight years in a university studying everything in the heavens and on earth except how to live, and if he doesn't kill himself before he finishes a course at college, he frequently does so in a few years afterwards. Here is another curious fact, a mother will sacrifice her life for the welfare of her child, but before it was born, she did not think it worth while to endow such vigor and character on her babe as to make it *fit to live*, and though she may love her infant babe far beyond any feeling that could be suggested by words, *the chances are one to five that she will kill it before it is a year old by improper feeding.*"

"Then you are a believer in the scriptural text that the iniquity of parents shall be visited unto the third and fourth generations."

"Yes, in a measure that is true, but not absolutely; that is, not all iniquities are transmitted. Nature constantly strives to correct the mistakes which injure. Were it otherwise, the weaknesses and vices continually taken up by each generation would soon extinguish the race, if none were cut off."

"What is the chief factor in producing the physical and moral imperfections of the race, doctor?"

"Well, part of our present social evils are no doubt due to false economic conditions, but if every individual was born right and properly educated even these would disappear."

"But as things now exist there must be other great factors besides economic ones that affect the individual."

"Yes, many things affect his existence, such as exercise, ventilation, sanitation, clothing, and each are so important, that thousands of lives are annually sacrificed because the natural laws of which they are a part, are violated; but while these affect many, the most important thing of all is food; it affects the whole world."

"Since you speak of it, I realize the force of what you say, for I asked a teacher about the quantity and proper proportions of the ordinary foods that would be required for health and vigor and he couldn't tell. He said the physiologies and books of hygiene only gave a little general knowledge, with very little practical information."

"I am not surprised that a teacher couldn't," said the doctor, "a great many physicians could not do it, for they are not employed to keep people well, but to drug them when they are sick, and so long as people prefer to pay for taking medicine, the doctors are powerless and unable to do anything better, however much they might desire it; but the doctor of the future will be employed mainly to prevent disease."

"Doctor, since you have said what you have, the question occurs to me, how do people live at all?"

"By mere accident or chance. They eat what they want, that is, what their appetite craves, or what may be

offered them, no matter whether it suits their requirements or not. If it makes them suddenly sick, there is
not much danger, but if their food is wrong for a number
of years, and if its evil effects are not quick in manifesting themselves, the doctor will finally have a much
more serious case to cure, if indeed a cure is not beyond
his power."

"Then, if I understand you, the people live almost
universally in a haphazard way and if they get sick,
rely on nostrums and doctors to cure them."

"Exactly so."

"How do you account for it?"

"It is partly due to the fact that people believe
that proper living is galling; that all the pleasures of life
would be cut off if they had to live by rule; but probably a far greater number are under the impression
that their work the weather, or natural causes, produce
their ills, when in fact they are self-inflicted."

"Well, I have heard these reasons so continuously
that I almost come to pity those people who are always
saying that something or other in their lives, either their
work, the weather, or some accidental circumstance made
them ill and wretched."

"Is there nothing in this?"

"Not much. Most persons can eat almost any food in
proper quantity at a proper time if properly prepared, and
as to weather and work killing people, who live in accord
with natural law, such would be as hard to find as a
dishonest alderman" said the doctor, with a somewhat
significant twinkle in his eye, then continuing, "it is no
harder to live properly than it is to speak or write grammatically; one doesn't have to think of all the inflections
of every part of speech in writing, for correct use of language comes by knowledge and practice, and good usage

is only difficult to the illiterate.   Just so as to living.   If
you really understand foods and their relation to life, it is
easy to be well.''

"But people often say that they don't live up to
what they already know, and what is the use of learning
more?''

"People who say that it is no use to learn because
they do not live up to their knowledge are unconscious of
their own ignorance.   It is true that no one applies all he
knows to each act of his life, and this fact is illustrated in
our daily conversation, for however well educated, few
persons speak correctly at all times; but would any one
say, that because of this being a fact that it is useless for
one to get an education?

"Nobody but an idiot would say so.''

"Here is another fact, few live up to their moral enlightenment, but according to the reason urged for not
learning more about our bodily existence, all the efforts
put forth to christianize and enlighten the world are useless.
We might go still farther and say, that but few people do
business as systematically as they know how; is a business education, therefore, of no use?''

"Doctor that is well put, and emphasizes the importance of training.''

"Yes, knowledge is the main spring of action and the
people who will not be controlled are those who are suffering from some defect the result of their own or others'
violation of physical laws.''   .

"Then you think people who have right knowledge
of living will not go far wrong, if they are not already
badly warped by somebody's transgression?''

"They will not, for it is absolutely certain that under no circumstance will the people go as far in their violation of what they know will injure them, even though

their inclinations lead them against their knowledge, as they would certainly do, if they were entirely ignorant of the effects of such imprudence."

"Then knowledge is useful to recover from errors in living?"

"Yes, when we are ignorant we not only injure ourselves, but not knowing the cause of the injury, we are likely to continue until we are beyond any remedy. *No language can sufficiently emphasize the fact, that there is nothing of so much value to us, as knowing how to live, and to know how and what to eat comes first.*

"Are we to understand that all the ailments of life come from improper food?"

"No, not all of them, but most of them do. Some come from hereditary tendencies, some are thrust upon us, such as infectious diseases, but if people were to eat the right kind of food in proper quantity, and properly prepared, sickness would scarcely be known at all."

"Has physical weakness much to do with our career as individuals?"

"Yes, everything; it retards moral and intellectual development, causes a craving for stimulants, drives people to crime, makes labor a burden when it should be a pleasure, causes life to be partially or wholly a failure, and frequently makes the individual a burden to society instead of a blessing. All of which are forcibly illustrated by our penal institutions, alms houses and asylums."

"There is no doubt, doctor, but what the people need enlightenment upon the subject of proper living more than any other?"

"Yes, unless it be the question of heredity and prenatal influences, but as you want to deal with questions for immediate results there is no field which could possibly offer you a greater opportunity for labor."

"What would you suggest as a proper scope for a book that would, in your opinion, be of most benefit to the people?"

"The qualities and properties of everything used as food should be given, and the best methods of preparing each food product. It would also be well to point out the deficiency, if any, of each food and what would be suitable under different conditions to make a complete diet. It would be advisable to mention those foods that have particular value as remedial agents, and suitable diet in all diseases."

"Whom should we urge to study a book of this kind?"

"I suppose most people would say that those who are ill need it most, and while it is urgent for them, the greatest good can be done by interesting those who are as yet too young to have suffered irreparable ill from bad example. *As this is beyond question the most important of all education it should be taken up and taught in our public schools as the most essential branch of the school course.*"

"Then you don't regard school physiologies as of much practical benefit?"

"As to that, it certainly isn't objectionable to study physiology and hygiene, but the only trouble is that much of it has no direct bearing on living, and *too* many suppose they are *well* informed when they have not learned any more about living than they would have done about house building, by simply examining a house and finding that it was made of brick, mortar, stone, wood and metal."

"That will shock some of the teachers."

"I hope not, for I was a teacher myself and studied and taught from the school physiologies, but what I did not know would have filled a large book. Too much attention connot be given to this because *experience teaches that those who are now healthy are gradually perhaps imperceptibly breaking their natural vigor,* so that with them it is only a question of time before they will have the common ailments with which everyone is familiar. If these can be reached they can be saved much distress, while those who are diseased and broken might not be worth but little after you have taken away all the causes which afflict them. It is a good plan to help all the people you can, but do not devote all your energy towards working over spoiled material."

"Then you think it better to save the coming and future generations."

"Yes I would seek them but let the others seek me.'

CHAPTER II.

DIGESTIVE ORGANS,
AND THE PROCESSES OF DIGESTION.

"Doctor, in beginning the study of any subject, it is of course very important to start right."

"That is true, and if we are to understand the source of health as well as disease, we must know something about the digestive organs and how they work to keep us well, and under what circumstances they will not, or can not work, and thus allow us to get sick."

"A great many people don't understand what is meant by digestion."

"Digestion is the process by which the various particles of food we eat are dissolved and changed by the digestive secretions and processes into suitable elements for the various uses of the body."

"Are the particles taken into and absorbed by the system made very fine?"

"Yes, the particles that are absorbed are so fine that they must be magnified several hundred times before they can be seen by the naked eye."

"This is very interesting, Doctor, where does the process begin?"

"It commences where a great many kinds of trouble begin."

"That must be in the mouth?"

"Yes, in the mouth and in the kitchen, and unfortunately, most people in this, as in other things, use their mouths and kitchens much but not well."

"What do you mean by that?"

"That they talk without thinking, eat without chewing, cook without knowing how, and eat more than they eat properly."

"That is because they don't know how to do any better?"

"That's a charitable view and no doubt true in part."

"At any rate one is astonished, at how little people know about living and that is true even of the educated classes."

"Yes it is. A good many people would dispense with their mouths for eating if they could, and shovel their food into their stomachs just as they would load a wagon with hay. When they get sick, they charge it to anything or everything except their own folly."

"Then the great fault is in eating too fast?"

"Well, as already stated, digestion is first of all a process of dissolving, and a good many people treat their stomachs as though they had better teeth in them than in their mouths. It is time for people to learn that they only have one set of teeth, and that if they continually impose on their stomachs, by compelling them to do the work that should be done by the teeth, sooner or later, their stomachs will get stubborn and not work at all."

"Yes, Doctor, but you forget, don't you, that many kinds of food are ground before they reach the mouth?"

"That would seem to be a good point, but somehow the creator of man did not anticipate mills, and consequently, arranged an important process of digestion in connection with the uses of the teeth, which cannot be avoided without positive injury."

"Then there is no safe way of cheating the teeth out of their grinding business?'

"None whatever."

"What is the important process that you have just mentioned?"

"No doubt, you have noticed that when you chew anything, your mouth is soon filled with a slippery ropy fluid, usually known as saliva."

"Yes, where does it come from?"

"It is a secretion that comes from glands within and adjoining the mouth each of which has a tube draining into the mouth."

"Have these glands names."

"Yes, the principal ones are known as parotid, sub-maxillary and sublingual glands and there are small glands scattered through the lining membranes of the mouth and tongue. These are called buccal (mucous and serous) glands."

"Do they all secrete the same kind of fluid?"

"Well, it is all a digestive agent, though the character of the secretion of each is different?"

What are the various uses or saliva?"

"It was formerly supposed that the saliva had no other use than to moisten the food, and no doubt every one has noticed that as soon as they commence to chew anything, the saliva commences to flow; for that reason, it appeared that the saliva was only intended to make the food soft so it could be swallowed easily, but with the aid of modern chemistry, we have learned that saliva is a digestive agent, which must be mixed with the food during the grinding of the same by the teeth."

"What is the nature of the secretion?"

"It is an alkaline solvent that dissolves that part of the food known as starch, gum, pectose and similar substances."

"In what is an alkali different from an acid?"

"Probably, the nearest we could describe it would be to say that it is the opposite of acid. If we mix them in proper proportions, according to the strength of each, both will become inert."

"Then digestion of all foods containing starch begins in the mouth?"

"It begins there if the saliva be mixed with the food but the fact that so many people swallow their food without chewing it, especially all soft foods, such as warm bread, mashed potatoes, pudding, oatmeal and all similar foods, there is not ordinarily sufficient saliva added to digest any quantity worthy of notice."

"Then the old saw, 'who eats slowly lives long' must be true."

"It is."

"Has the saliva any effect on foods other than the starches?"

"Not as a digestive agent, but it aids in keeping the particles of food that are crushed by the teeth from adhering together."

"How much saliva is ordinarily secreted in a day?"

"Those who have carefully estimated it, say that eight to ten ounces are daily secreted."

"That would hardly include tobacco and gum chewers, would it?"

"No, chewing tobacco is a perverted use, and tobacco chewers have saliva with which to bathe a considerable portion of the earth but very little for their food."

"Of what temperature does the saliva act on starch?"

"At 103° to 112° F. It does not act below 85° F. to any extent, nor over 168° F."

"Then moderate temperature is an important thing in digestion?"

"Yes, this explains part of the ill effects of ices and very hot drinks."

"What is the other part?"

"Direct damage to the mucous membrane."

"Is there anything else about the mouth that aids digestion?"

"Nothing that aids it, but something that doesn't aid it."

"What is that?"

"Filth.   Some people keep their mouths like garbage boxes.   They allow all kinds of food to lodge and decay until it even rots their teeth, and then they have a mouth tainted with decaying food and decomposing bones, which is a harbor for the various kinds of bacteria."

"What harm does this do?"

"When food is eaten, these foul accumulations and bacteria are carried to the stomach, and no doubt are often great factors in disturbing the stomach and general system, and one of the sights calculated to make one pity the human race is to see persons cut holes in their flesh to make themselves beautiful with jewelry and yet carry a mouth and teeth coated with putrid matter so offensive in odor that it is disagreeable to be near them."

"What becomes of food when it leaves the mouth?"

"It passes down a tube called the oesophagus (gullet) into the stomach."

"Do people understand how their stomachs are constructed?"

"No, a great many people suppose their stomachs are copper lined, or at least their habits lead one to that conclusion."

"Why do you say that?"

"Because they have no regard for their stomachs and give themselves no concern as to the character or quantity of what they put in them."

"In what particular?"

"It is not an uncommon thing to see people eat soup scalding hot and then drink ice water to cool it. Others make a catch basin of their stomachs and pour in several gallons of beer or large quantities of stronger liquors."

"There are but few who do not use mustard, pepper, horseradish and other intense irritants, while those who are continually taking poisonous drugs are legion. This is not all, the stomach is not supposed to rebel no matter how coarse or tough the food. nor how incompatible the mixtures that ignorance pours into it, and as a result of all this, if the aches, pains, diseases, misery and deaths could be measured by volume they would make a pyramid from the earth to Jupiter."

"Suppose you tell us something about the stomach, doctor?"

"I can do that best by first showing you a photograph of it. (See page 20 for illustration.) It is generally described as an irregular shaped sack or pouch, and will hold in normal condition from two up to three and a half pints, although in one case the stomach of a grown person was known to hold only a half pint. Abnormal size is very common, because the majority of the people use their stomachs as a receptacle for the most outlandish collection of indigestible material which a pampered civilization can supply. This stretches them so that they are made to retain several gallons of liquid and food under which the system groans with the weight of its torture. The modern stomach exposed to view looks much like a fourth of July balloon. The inside of the stomach is lined with mucous membrane, very similar to that of the mouth. This is arranged in many folds running lengthwise. If the membrane be examined by a microscope, innumerable pits are seen. These pits indicate the presence of gastric glands."

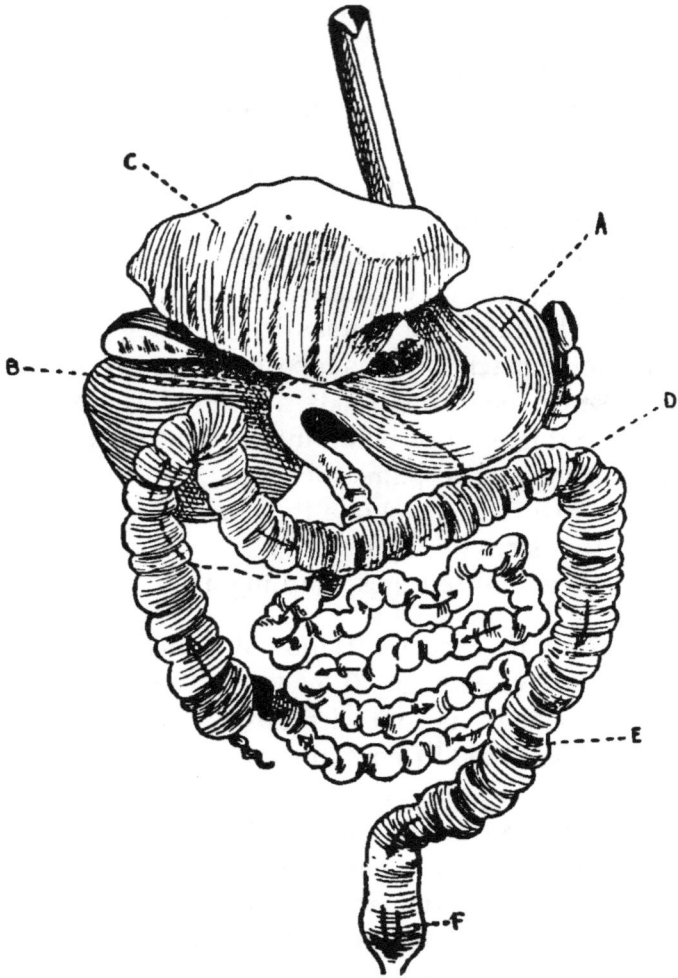

Fig I.

A, stomach.   B, pyloric end of stomach.   C, liver turned
up to expose stomach.   D, large   intestine.   E, rectum.   F, annus.

"Do the gastric glands perform an important part in digestion?"

"They do, for they secrete what is commonly known as gastric juice."

"Are there other glands in the stomach?"

"Yes, mucous-forming cells that secrete mucus."

"Is the gastric juice anything like saliva?"

"Not in the least, for the gastric juice is acid and the saliva alkaline."

"Then it is the acid that dissolves the food?"

"That depends upon the kind of food you mean. Properly speaking, it is not the acid alone, but the secretion of acid and a substance called pepsin, acting together, that dissolves tissue forming foods, but not starch. There is another substance in the gastric juice called rennet. This is also called a milk-curdling ferment."

"How did they ever find out anything about what was in the stomach, and what goes on in the process of digestion?"

"Well, not much was known prior to 1822."

"What was discovered at that time?"

"That was the time when a man by the name of Alexis Saint Martin had his stomach accidently torn open by the discharge of a musket."

"What was the wound like?"

"The front part of the sixth rib was blown away, the lung and diaphragm torn; but after a long convalesence he recovered except that there was a large fistulous opening into the stomach. This at first had to be bandaged, but after a time a portion of the mucous membrane of the stomach prolapsed until it hung down over the opening, thus acting as a sort of a curtain to the stomach."

"That was remarkable."

"Yes, it furnished Dr. Beaumont who treated Saint Martin, a practical method of observing the process of digestion.

"Who was Dr. Beaumont?"

"A surgeon in the service of the United States."

"How did he describe the digestive process?"

"Dr. Beaumont, in writing of his observation on the stomach of Saint Martin, states that when food first enters the stomach the movements of the stomach are feeble and light, but as digestion goes on, they become more and more vigorous, until the action of the stomach thoroughly churns the contents within it. The food travels from the upper opening along the lower or greater curvature, to the pylorus, (the end where the food is discharged into the intestines) returns by the upper or lesser curvature, while at the same time the movements of the stomach turn its contents inward so that every particle of food in the stomach comes in contact with the freshly secreted gastric juice. As digestion proceeds, the contents of the stomach becomes more and more acid, and the contracting force of the stomach becomes greater, so that it constantly throws its contents inward from its own walls as well as downward towards the opening into the small intestines."

"When does the contents of the stomach pass out?"

"Under normal conditions, some of it passes out, or rather is ejected, as soon as it becomes sufficiently liquefied. Just what governs the expulsion of the food from the stomach is rather difficult to determine; it is not merely the fact of its becoming a liquid, as water, no matter what its temperature, remains in the stomach several minutes and is then discharged into the intestines, where absorption takes place. It also frequently happens that solid food is not dissolved at all in the stomach, and if the irritation is not great enough to cause vomiting, it passes into the intestines, but just at what time or condition, has not been determined. It appears that under some conditions, solids readily pass out of the stomach, while in others liquids remain a long time, so that the discharge of food from the stomach is not entirely a question of liquefaction, (i. e., becoming a watery liquid.) The ordinary length of time which water remains in the stomach when there is litttle or no food in it, is about 15 minutes, but it may remain hours, when the activity of the stomach walls is impaired."

"Does the stomach always empty itself?"

"It should, but sometimes it must be vomited to do so. Food has been known to remain in the stomach several days and then be vomited."

"Of what use is this knowledge, doctor?"

"Well, I only wanted to explain that part of one meal may remain in the stomach undigested until it is time to eat another."

"What will then happen?"

"After several hours, if the food is not digested, decay will set in, and when one meal disagrees it may not be noticeable. but the decaying portion remaining in the stomach will almost certainly destroy the digestion of the next one; whereas, had the stomach been empty, digestion would readily have taken place. This is the reason why it is more or less difficult to tell what agrees and what disagrees with us, because the undigested meal may not be noticeable until another meal is added, or not at all until symptoms of sickness occur."

"Doctor, you stated that the stomach did sometimes permit undissolved food to pass out into the intestines, does any harm come from it?"

"Yes, great harm, and such ailments as cramps, colic, diarrhoea and catarrh of the gall bladder, causing gall stones, are common results from coarse substances passing through the stomach into the bowels."

"Has there been any extensive investigation made about digestion since that of Dr. Beaumont?"

"Yes, investigations have been going on almost constantly ever since, and have been much aided by what we might properly term Modern Chemistry. A German physiologist in 1831, discovered that saliva digests starch, i. e., turns it into grape sugar, sometimes termed maltose. Since then, repeated experiments have been made upon man and animals. Thousands of tests have been made by siphoning the gastric juice and partly digested food from the stomach, and also almost every conceivable test has been made on dogs and other animals."

"Doctor, you say saliva is alkaline while the gastric juice is acid. A while ago you stated that these were antagonistic and that the alkali neutralized the acid. Now, how can digestion be carried on by two elements directly opposite to each other, one neutralizing or destroying the effect of the other?"

"That is a good question, and a right nnderstanding of the answer would clear up many of the doubts and difficulties concerning food, or rather our diet. In the first place, it would be well to remember that the saliva makes its appearance in the mouth; that it has great effect on the digestion of starch, or starchy food; that it should be mixed with the starchy foods as thoroughly as possible. Then, when the food reaches the stomach, the gastric juice only begins to flow. It will thus be seen that there is considerable time for the digestion of the starch before any quantity of gastric juice has been secreted in the stomach, or to put it another way, the starch digestion begins in the mouth and continues after reaching the stomach until the stomach has secreted a sufficient quantity of the acid gastric juice to counteract the effect of the alkaline saliva which the food received in the mouth. Ordinarily, it would require from 15 minutes to a half hour for the stomach to become sufficiently acid to neutralize the amount of saliva that ought to be mixed with the food during its mastication before it reaches the stomach. As the stomach gradually becomes more and more acid, the starch digestion gradually lessens until it entirely ceases. Then the action of the stomach walls becomes quite intense, and gastric digestion properly begins."

"What do you mean by gastric digestion?"

"I have tried to make it plain that the saliva has no solvent action upon proteid or tissue forming foods. It is this class of foods that are dissolved or at least should be, in the stomach, by the secretions therein."

"What class of food do you call starches?"

"Generally speaking, all the vegetables with the possible exception of peas and beans, are essentially starch, and even peas and beans contain a per cent of that substance. The foods acted upon in the stomach are lean meats of every kind, eggs, milk, cheese, fish, and the vegetable casein in peas and beans and the gluten found in wheat and in other cereals."

"Is digestion completed in the stomach?"

"Not by any means. By far the most important part takes place below the stomach—in the small intestines."

"Then, according to the statement you make, the stomach is not of much use, and knowing something of the trouble and pain it gives it looks as though we were constructed on immature plans."

"Not at all. The stomach has its use and a most important one, although Czerney in 1876, at Heidelburg, Germany, removed the entire stomach of two dogs. No mention is made as to the effect on one of the dogs, but the other lived from 1876 to 1882, when he was killed for the purpose of making an examination as to his condition. At the time his stomach was removed, the dog weighed 5,850 grams (22 lbs.), a month after he weighed considerably less, but during the year his weight increased to 7,000 grams (29 lbs.)"

"Has any person ever lived with the stomach removed?"

"There is no such case on record, although the pyloric end of the stomach, (i. e. the end on right side) has been cut out and the intestines sewed to the stomach. Such operations have seldom been successful, but it is probably due to the fact that they have never been made until the patient was almost dead of some malignant disease, such as cancer or ulcer."

"You have made no mention, Doctor, of fats. What action has the gastric juice or saliva on them?"

"Until within a few years it was supposed that the gastric juice had no effect whatever upon the fats, but modern investigation has changed that view somewhat, and it is now understood that the gastric juice is capable of breaking down or disintegrating fat cells thereby setting the fat particles or globules free. This, no doubt, is a great aid to intestinal digestion. It is also believed that the fat is to some extent changed into fatty acids and glycerine by the gastric juice."

"Has the gastric juice any other action?"

"It converts cane sugar into grape sugar, thus preparing it for absorption into the system."

"Is there anything besides starch that is not greatly acted upon by the gastric juice?"

"Yes, cellulose."

"Where is it digested?"

"It is digested somewhere in the apparatus of the lower animals, but nowhere in man; in fact, it keeps starch from digesting, because starch is encased in small cellulose cells, and unless the cells are ruptured by cooking or by mastication, starchy cereals and vegetables are almost wholly indigestible."

"How long does food ordinarily remain in the stomach?"

"From one to four hours, frequently longer."

"What are the modifying conditions?"

"Much depends on the kind of food, upon the cooking, and the mixture of different kinds of foods."

"Doctor, that is not plain to me, will you give examples?"

"Well, meat and tough vegetables, like peas and beans, require longer time for digestion than something that is easily dissolved, like the white of an egg. Then as to cooking, the longer meats are cooked, especially if roasted or fried, the harder and more insoluble they become, as heat coagulates, that is, makes the albumen in meat more solid."

"Is there anything else that makes meat difficult o. digestion?"

"Yes, being saturated with fat, because the gastric juice of the stomach has only a limited effect on fat, and if eggs or lean meat be fried or saturated with it, the particles might aptly be termed encased, and could only be acted on to a limited extent, if at all, by the digestive agents of the stomach. This is the reason why fried lean meat is so hard to digest."

"Is this all that determines the period of digestion?"

"No, there are many other things. The fineness of the particles of food has much to do with it, and it will not require any labor to demonstrate that a particle, say the size of a pea or bean would not be so quickly dissolved, if it be dissolved at all, as a particle as small as very fine flour, so that the length of time food should require for digestion depends much upon how finely it is masticated or artificially divided, and this applies equally to both meats and starches. Another factor is the amount of acid in the stomach."

"How does that affect the duration of digestion?"

"Well, some persons secrete very little acid, and are almost wholly unable to digest meats; others have such strong acid secretions that they digest meats very quickly, but that very fact might in a measure prevent starchy foods from being dissolved by the saliva, so that the kind of food and the amount of acid in the stomach are both elements affecting the period of digestion."

"Is this all, doctor?"

"No; perhaps one of the most important of all is the demand of the system for food."

"How does this affect digestion?"

"Well, if the system has previously been supplied with more food than it can use, nature has some way of protecting herself by not adding to the burden already carried. Of course, if the intestines are loaded with matter and their action slow, the food would not be quickly drawn downward. It is believed that when the system is clogged or there is an excessive accumulation of matter in the bowels, that the stomach must necessarily be in sympathy, and it sometimes happens that foods remaining too long in the stomach and decaying there is the first symptom pointing to the fact that the digestive organs have been overloaded and that there is no demand for food."

"Some people say that the amount of drinks or fluid taken into the stomach has much to do with the duration of digestion."

"That is true. If the digestive juices are greatly diluted they must necessarily be much less active than if they have their full strength."

"What about the temperature of the fluids taken into the stomach?"

"It also influences digestion, from the fact that the temperature of the stomach must be maintained at about the normal heat of the body, If cold drinks be poured into the stomach, as a matter of course, digestion will be delayed until the stomach can be re-warmed."

"Do individual peculiarities have much to do with the time required for digestion of food?"

"Yes, some people have very active stomachs but yet have inherited some antagonistic tendency to certain foods"

"I have often heard people say that when people are in serious trouble that they were likely to suffer from indigestion, why is this?"

"Well, anything which affects the nervous system and in that way disturbs circulation, will affect digestion."

"What is the theory of this, Doctor?"

"It is because the stomach requires a large supply of blood, and if the blood from any cause is in excess in other organs the supply of the stomach will necessarily be diminished. Great mental excitement keeps the flow of blood to the head instead of the stomach, and the same may be said of every vigorous exercise. There is still

another cause for the various periods required for digesting the different foods, that is, their chemical effect on
each other.   To illustrate, tea contains a large amount of
tanic acid.   If strong tea should be drunk after eating the
white of eggs, the tanic acid of the tea would precipitate
the albumen of the eggs and make it entirely indigestible.
This is about the same process as that of tanning leather"

"Then you don't attach much importance to statements
that certain articles of food are digestible in a certain
time?"

"No, although something like an ordinary average
might be estimated, for instance, well-done meat should
ordinarily be digested in four or five hours, or six at the
most, although sometimes it is never digested.   Meat
properly cooked should be digested in about 3 hours, and
experiments with raw meat show that under fair conditions, it will be digested in 2½ hours."

"Then according to that, cooking meat makes it more
indigestible?"

"As a general rule, it does, and as it is ordinarily cooked, it makes it much more so.   The same may be said of
eggs.   Raw eggs could be digested in about two hours;
hard fried eggs, if at all, in four to six hours."

"How about vegetables?"

"Peas and beans being very tough require three to
four hours.   Ordinary bread, if good, about 2½ hours."

"Do liquids require much time for digestion?"

"Water or fats and oils taken on an empty stomach
would not ordinarily remain in the stomach but a few
minutes."

"How about milk?"

"Milk is taken as a liquid but it becomes a semi-solid in
the stomach, and requires one-half to two hours for

digestion. Of course, as already explained, these estimates are only mere outlines which are varied by many circumstances."

"When digestion begins, does the food leave the stomach as fast as digested?"

"No, although at intervals small amounts of dissolved food are ejected from the stomach," but the greater portion of it remains in the stomach until digestion has been sufficiently completed to allow the food to pass into the intestines."

"Then according to your explanation, the stomach is a sort of a reservoir, in which the food is prepared for further changes in the intestines."

"Yes, it might be called something of a dissolving vat."

"Why is this?"

"Well, the intestines are much more susceptible to foreign substances than the stomach."

"I don't understand what you mean?"

"I will explain; the stomach is an organ of considerable size, while the intestines have smaller diameter and greater length."

"About how long?"

' The small intestines about 20 feet or more, the large intestines about 5 feet. Where the intestines join the stomach is called pylorus. Where it joins the large intestines, ileo-caecal valve."

"What is their general structure like?" '

"It is a small tube containing muscular layers running lengthwise and also around the intestines. The blood vessels and glands are very numerous. The inside of the intestine being lined with a mucous membrane similar to that of the stomach, but in the stomach the folds run lengthwise while in the intestines they are crosswise."

"What is the principal agent of intestinal digestion?"

"Pancreatic juice, which is secreted by the pancreas."

"Then according to that, the pancreas is the most important of the digestive organs?"

"Well, the digestive organs act as a unit, each being essential, although the pancreas furnishes the most indispensable part of the digestive fluids, because digestion can go on in the intestines if the food be fine enough, even though there be no preparation made in the stomach, or by the mouth.

"What kind of an organ is the pancreas?"

"It is a long, narrow gland of reddish cream color, but of course the color varies according to circumstances."

"Where is it located?"

"It lies behind the stomach in the rear wall of the abdomen."

"In what way is it connected with other organs?"

"It has two tubes or ducts, emptying into the intestines three or four inches below the lower end of the stomach."

"Doctor, you haven't explained the general character of the pancreatic juice?"

"It is an alkaline fluid containing many chemical elements."

"How do these elements act in furthering the processes of digestion."

"The pancreatic juice has three distinct properties. It dissolves all preteid foods, such as meat and eggs, also has a very active solvent, which quickly digests starch and it has still another element which splits up or decomposes the fats, splitting them up into extremely small particles making a creamy substance closely resembling soap."

"Is there any other digestive agent besides what you have already mentioned?"

"Yes, there are others. The one most universally known but probably the least important, is bile."

"I have often heard people speak of having bile on the stomach, is this true?"

"Not ordinarily at least, the bile duct from the liver or gall bladder empties into the intestines several inches below the stomach, and it is only when the proceedings of nature are reversed as in cases of extreme vomiting, that the bile is brought up through the stomach."

"Of what use is bile in digestion?"

"The uses of the bile are still a subject of more or less dispute, although it is generally understood that the bile is a very important factor, in connection with the pancreatic juice, in preparing fats for absorption. A number of experiments have been made upon the digestion of animals without bile. and it was found that a large per cent of the fats were not absorbed. This is said to be true also in jaundice where the flow of bile is obstructed or in some way deficient."

"Has it any other uses?"

"Being strongly alkaline, it arrests the action of the stomach juices and aids in preparing the food as it comes from the stomach for pancreatic digestion; this being an entirely different process from that carried on in the stomach. Bile will dissolve small quantities of fats, and has long been used to remove grease stains from delicately colored fabrics, but its action alone without the pancreatic juice is not very marked."

"What other uses has the bile?"

"It is claimed that it will to a certain extent prevent abnormal fermentations or decay of the food in the intestines and there is no doubt but what bile acts as a laxative in the bowels. It also acts as an antidote to poison known as nicotine which is one of the active principles of tobacco. Numerous experiments have demonstrated the fact that it is about the only known substance which increases the flow of bile, although various drugs in a measure accomplish the same result by setting up an activity of the bowels; these are known as cathartics."

"Is there any other secretions found in the intestines that affect digestion?"

"There are numerous small glands throughout the intestinal canal. These secrete an alkaline fluid but so far as has yet been determined they have no other use except to convert starches into sugar and perhaps aid in keeping the contents of the intestines from becoming excessively acid through fermentative processes."

"I don't see what is to be gained about all this talk about what goes in the intestines?"

"Then you don't care to know how to keep alive. That is why a good many people don't live, they merely exist, at least they must have constant assistance from their doctor."

"Then, what is to be learned by this?"

"First, that there is a limit to the size of the lumps— the coarseness of the food that can be properly disposed of by the stomach."

"Suppose this is violated, what is the effect?"

"Very likely cramps or inflammation which will probably cause serious injury and even death. The second thing to be taken notice of is that 25 feet of intestines require something to incite their action; i. e., waste matter sufficient to give them something to do."

"Suppose the diet does not furnish the necessary waste what will be the result?"

"It would seem from the construction of the intestines having folds almost their entire length, that it would be difficult to get anything through them. Can you explain how this is accomplished?"

"By activity—peristaltic movement."

"What is that?"

"The peristaltic movement of the intestines is a wave-like movement similar to that of a caterpillar in motion."

"I perceive if there be so much movement there must be freedom?"

"Now you have struck a great point. Tight waist bands and tight corsets hinder peristaltic action of the intestines, and the man or woman who reduces the size of a natural waist (which a very large per cent of women do) deserves to be called an artist with more vanity than sense."

"Does any digestion take place in the large intestines?"

"Not in the sense that it does in the small ones. The processes of the large intestines are those of decay and it is believed that particles of food that have not previously been acted upon are to some extent dissolved by the action of bacteria, a fermentative process."

"How is the digested food taken up by the system?"

"The entire length of the intestines contain little tongue-like projections called villi, which are attached to the folds of the mucous membrane. These take up the digested particles by a process called osmosis or absorption from without, and they are carried into the circulation."

"Do they immediately become blood?"

"Some portion of the food so absorbed immediately

enters into circulation as part of the blood, while other parts enter either the lymphatics or lymph glands or portal vein, and carried to the liver and probably modified to some extent by that organ from which it is taken up as needed."

"Doctor, will you kindly sum up the important things to remember about digestion."

"First, food must be properly prepared; second, it must be thoroughly masticated, ground fine and thoroughly mixed with saliva, especially if it contain starch; third, no fresh food should be taken into the stomach during the period of digestion; fourth, food should be properly proportioned, containing the different elements required for the purpose of sustaining life."

"Doctor, you have not mentioned how often one should eat?"

"That is somewhat a matter of habit. The savage tribes eat when they are hungry or when they can get food."

"Would you advise people to be guided by their appetites?"

"Not by any means. It is better to have fixed habits, although if there be cause for hunger and need for food, this feeling should be gratified within reasonable limits."

"What do you mean by cause for hunger?"

"Hunger may be either normal or abnormal, that is, it may come because one eats but little food, and takes a large amount of exercise, while abnormal hunger, which is even a more intense craving for food, results from disease, or excessive stimulants such as condiments or alcohol."

"How can one tell whether the appetite is normal or abnormal?"

"By amount of food eaten and amount of exercise taken."

"How often then should one eat?"

"That is difficult to say, for it depends on habit, ability to digest food and the activity of the person."

"Then a uniform practice of eating three times a day is not always best."

"No, many persons would remain in better health when eating four or five times daily, but ordinarily three meals a day are sufficient, and some even claim that two meals agree better than three. This is especially true of brain workers. The two meals should be at the beginning and end of the day."

"What class of persons should eat more than three times a day?"

"Persons of very weak digestion as convalescents from acute diseases, or people who are very fat."

"Won't this have the tendency to make them take too much food?"

"On the contrary, the inclination is to take much less. Weak stomachs need food in very small quantities, and eating often satisfies the appetite."

"Can you give general rules?"

"Yes, no one in active labor should go longer than six hours without food."

"You say appetite should be considered, in what way?"

"In this way, if you were to eat a very light breakfast at six or seven o'clock in the morning and have active exercise, it would not be unnatural to be hungry at 10, and it would then be better to eat something than to wait until 12 for the regular meal."

"I have always heard that it is bad practice to eat between meals, now you advise that under some circumstances it be done."

"The objection to eating between meals is not well understood. What is meant by the general outcry against it, is that no food ought to be taken into the stomach while what has been previously eaten is in process of digestion."

"O, I see, eating between meals if the stomach is not empty, is after all a bad practice?"

"Yes, it is very bad, for it keeps food in the stomach too long and very likely causes it to decay, because part of the amount previously eaten will probably be retained until the fresh food has been added. This necessitates the retention of the previous meal until the second is digested, and therefore causes increased delay. The practice cannot be too strongly condemned."

"Doctor, I am still confused. You say that a light breakfast at six or seven and active exercise might make it proper to take food at ten, what would you do about the regular meal if it came at 12?"

' This is a matter which requires judgement. In such a case, a ten o'clock lunch should consist of some fruit that is easily dissolved, like a baked apple. If good fruit cannot be had, then a little milk, sugar, or even bread in small quantities."

"Then the habit of eating meat, pickles and beer lunch is objectionable?"

"Extremely so; no liquor (if it is to be drunk at all) should be taken on an empty stomach, but to do so and to eat pickles and salads besides, is a species of folly so great that it is difficult to understand how a rational person can do it."

"What should be the principal meal of the day?"

"Well, for most people the principal meal should be in the middle of the day, although breakfast may be a heavy meal, if not convenient to eat anything but a lunch in the middle of the day.   The evening meal should always be the lightest, because the system is most relaxed and the least capable of digestion."

"How about eating at bed-time?"

"If there has been active exercise and the hour for retiring late, a little food may be beneficial. To persons who have an inclination to insomnia (sleeplessness,) a little food will often be conducive to sleep and there is nothing we could more strongly recommend than Horlick's Malted Milk."

"Why does it make one sleep?"

"Taking a little food at bed-time has a tendency to draw the circulation from the head to the stomach, and whenever the excessive flow of blood to the head is diverted, then sleeplessness will be supplanted by restful sleep."

"I have always understood that eating food just before retiring had a tendency to keep one from sleeping at all?"

"There is some truth in this—depends upon the quantity and kind of food. A hearty meal always has a tendency to make one go to sleep, but if the meal is of such a character that it is a struggle to digest it, it almost naturally follows that the circulation will be disturbed much more than it ought to be; hence, the weird dreams and 'night mares' so called, are common incidents to late suppers of rich and indigestible food."

"What is the significance of sleepiness after meals?"

"Well, if there be great drowsiness after meals, it indicates either weak digestion or nervous exhaustion."

"How does exercise aid digestion?"

"Exercise aids by increasing the circulation and in that

way clearing the system of waste, and by burning up the food, thus creating a demand for a new supply.''

''Then the more exercise the better?''

''Not at all, exercise to the extent of great fatigue weakens very much and if such be unavoidable, it is much better to take some rest before eating and also after.''

''Doctor, I have noticed that some people soon become ill if they do not sleep enough, why is this?''

''Lack of sleep in some way disturbs the nervous system and weakens its tone. It follows then that inasmuch as every organ of the body is controlled by the nervous system, when it is disturbed every other organ will most likely be so.''

''Some contend that the use of tobacco aids digestion.''

''If it aids one, it hurts ten thousand, for it both depresses the action of the heart and affects the nervous system, and is therefore an unmitigated evil and universally injurious to all persons in normal condition, although it might be useful in some cases as a drug.''

CHAPTER III.

CLASSIFICATION OF FOODS.

"The food we consume serves us in two ways; first it supplies material for tissue and also for the bones; second, it furnishes us fuel for bodily warmth and action."

"What foods are required for these purposes?"

"While most writers divide foods into many classes, practically there are only two, that is, foods for building or repairing the body and foods for furnishing heat or force."

"Then you would only divide food into two classes."

"Yes, foods for building or repairing the body are called tissue-forming foods, they are also known by other names which are used to express the same thing."

"What are the names?"

"The most common name applied to tissue-forming food is the term proteid, or protein. Another term almost equally well-known is that of nitrogen or nitrogenous foods. Still another known as albumens or albumenoids. These various names are used interchangeably for the same purpose, and the reader should not be confused thereby."

"What foods belong to this class?"

"Lean meat, eggs, fish, milk and cheese are the foods most extensively known as tissue forming foods, but peas, beans, lentils and wheat gluten have a larger per cent of tissue-forming substances in proportion to their starch, than is ordinarily required for the human system. Properly speaking, they should be classed with tissue formers."

"What foods are known as fat or heat producers?"

41

"All fats and oils, starch, sugar, gum, pectose and waste material are all termed force producers. The foods belonging to the starchy class, including gums and waste material are usually termed carbo-hydrates, while the fats are known as hydro-carbons."

"In what classes of food do we find these different properties?"

"All the animal fats and oils, vegetable and fruit oils, sugar, starch and vegetables generally "

"Are there any foods that belong to both classes?"

"Yes, many of the foods in common use belong to both classes, that is, are both tissue formers and force producers. Milk, meat and eggs, all contain fat, and are therefore force producers by reason of the fat they contain, while the cereals, especially wheat and oats, contain nearly the proper proportion of tissue forming and heat producing substances. Ordinarily, the animal foods are called nitrogenous and the vegetables non-nitrogenous or heat producers."

"I understand that the system contains much mineral matter, that the bones are substantially all composed of it. Where does the supply come from?"

"The largest element of bone formation is lime, called calcium, while salt known as chloride of sodium, potash known as potassium, magnesia known as magnesium, and sulphur and iron and traces of other minerals, are found in various parts of the body. These various mineral elements are usually known as salts, or mineral matter, and exist in various compounds, generally known as chloride, carbonate and phosphate of sodium; chloride, carbonate, sulphate and phosphate of potassium; carbonate, sulphate and phosphate of magnesium; and phosphate of calcium."

"Are there any other uses for mineral salts, in the body, except for bone formation?"

"Yes, but it would be rather difficult to explain them to the laity."

"What is the use then of all this description then?"

"Simply to show the necessity of eating food that supplies these elements."

"Then it is a matter of great importance after all."

"Yes, many diseases result from not knowing this fact."

"What are some of them?"

"Rickets in children, anaemia, chlorosis, excessive growth and other ailments."

CHAPTER IV.

WATER THE PRINCIPAL ELEMENT

OF THE BODY.

"The human system is made up of many chemical elements, the principal part of which is water. The second largest element is carbon and next to it is nitrogen, while calcium is the largest element of the mineral substances. Those of less quantity are magnesium, sodium iron, sulphur and traces of other metals."

"Doctor, must our food contain all the elements of the body?"

"That is the conclusion from the experiments that have been made, although some of the elements are so small no demonstration has ever been attempted. It is but natural to conclude though, that if water, carbon and nitrogen and lime are indispensable that all the other elements would be so."

"I take it from what you say that water is the most important of all foods?"

"That is true. There is nothing taken into the system so indispensable as water, for it constitutes about 70 per ct of the weight of the body, and as the evaporation from the body is large, and being a vehicle for carrying off the waste and poisonous products of the system, more water is required than any other food, and if it is not frequently supplied, the blood would become too thick to circulate and death would result."

"Doctor, I suppose you have seen the statement in advertisements of liquors that water kills more people than whiskey?"

"Yes," said the doctor, "and strange as it may seem,

there is some probability that the statement is true; at least, it would be a good question for debate."

"You don't mean to say that people drink too much water, do you, Doctor?"

"As a rule, they don't. More drink too little than too much. It is not the quantity but the quality that kills, as the people of Hamburg learned in 1893 when they were scourged with the cholera."

"Is there much impure water?"

"Much! Why don't you ask whether there is any pure water, for such a question would be more in accord with the facts."

'What are some of the sources of polution?"

"Wells are polluted from surface water by spilling dirty water on its covering, by filtration from barnyards, privies, feed pens and street sewage. River and lake waters by sewage, decaying vegetable matter, and refuse of all sorts thrown into them, but this belongs to sanitation to be treated in a separate volume and we should not venture out too far."

"Cannot the people tell by seeing whether it is clear?"

"No, the clear sparkling water may be laden with death dealing impurities which may be vegetable or chemical, and may even have typhoid or cholera bacteria in it, but by boiling, it can be made wholesome and many impurities may be removed by various methods of filtering."

"But people object to boiled water, it tastes too vapid?"

"That can be easily overcome by agitating, like making milk shake or lemonade; in fact, the aeration these drinks get by being shaken is partly what makes them so pleasant."

"Can you give some suggestion as to how much water a person should take in twenty-four hours?"

"We take much of our water in what we call our solid foods; but unless we eat watery foods, like green fruits, the smallest requirements would be at least three pints daily, in addition to what would ordinarily be consumed in the foods including milk, tea and coffee. Of course, exercise, temperature and the size of the individual would all be varying circumstances and one would drink a great deal more water in very hot weather than in moderately cool or cold weather."

"Do people injure themselves by drinking water?"

"That is putting it mildly to say the least. Every tank and pitcher of ice water ought to be labeled with skull and cross bones."

"Why so?"

"Did you ever snow-ball? If you have, you have noticed that though your fingers would be for a time nearly frozen, after a while they would sting and burn with heat."

"Yes, I have noticed it. Is that the way ice water acts?"

"It is. When the blood returns, that has been driven away by the cold water, reaction takes place and if continued, the excessive flow of blood causes congestion, resulting in inflammation and general derangement of the stomach, usually known as ice water catarrh. It's a prolific source of what is generally known as summer complaint, indigestion, cramps, dull headaches and general languor, and dilatation of the stomach walls or enlargement of the stomach. There is another thing. Have you ever noticed if you take a large draught of ice water when you are very hot that you feel somewhat as if you were going to faint? Well, that results from the shock to the nerves in the blood vessels thereby affecting the activity of the heart and the fainting sensation is not caused by the heat as is generally supposed."

"That may be all very true, doctor, but think how cool it keeps one."

"Where did you learn that? That is one of the most absurd notions of the age. We get cool by what assists in radiation of heat, and whatever opens the pores of the skin and allows the heat to escape, makes the body cool. Trying to warm a tank of ice water in the stomach does not take away the heat from the system; but on the contrary prevents its escape from the body."

"Then you would advise the people not to drink ice water?"

"I would substitute cracked ice, which answers the place much better and is much more effective in rapidly lowering the temperature of the body. If ice be retained in the mouth until melted, the water becomes warm before it reaches the stomach. This entirely avoids the evil effect of ice water and prevents congestion with its train of ills."

"This does not give much water, does it doctor?"

"No, it is not the quanity of ice water that is drunk that gives such satisfaction, but the contact of cold to the mouth and throat, and melting ice seems to answer the demands of internal heat better than any other agent at our command."

"But cracked ice is not always easily obtained."

"Then it is better to only use part ice water and part of water much warmer, but people may drink ice water in very, very small quantities and very slowly. No one should drink more than a gill of very cold water at a time. No cold drink of any kind should be taken sooner than three hours after meals. In hot weather where there is an inclination to drink too much, a little water should be taken with greater frequency."

"Doctor, did I understand you to say that no cold water or cold drink should be taken soon after meals?"

"Well, not only should no cold water be taken, but no cold drink of any kind should be taken at meal time nor before digestion is completed."

"Why, I supposed the cold drinks would keep the food from souring too quickly and consequently be a good thing."

"You must have been reading about the animals of prehistoric races, said to have been preserved for several thousand years by being frozen. The fact is, the water cannot remain cold in the stomach, and if it did there would be no digestion and no use of taking food at all. As soon as the ice cold drinks are poured in, digestion stops until the temperature can be brought up to normal heat. This has a tendency to exhaust the working capacity of the stomach and if the cold drinks are repeated with great frequency, the stomach becomes permanently enlarged and digestion is paralyzed. This is one of the principal reasons why so many people are unwell and have diarrhoeas and lack force and energy in hot weather."

"Then cold drinks are worse in hot weather than in cold weather?"

"The principal is just the same, only the system is more ennervated in hot weather and the inclination to take cold drinks when we are warm is much greater than at other times; therefore, the injury from cold drinks is much more common in hot weather."

"I have noticed that many people drink hot water, while some say it is best to drink nothing at all during meals."

"Clear hot water is occasionally a useful agent for such ailments as result from acute indigestion and where there is no chronic enlargement of the stomach. Hot water drinking having been advocated originally for a few minor ailments, has been taken up by the multitude as a cure-all for every disease of the digestive organs. We can consequently call it a fad and a pernicious one at that. It is used ignorantly in many ailments where it acts as a direct irritant. Its most potently evil effects have been witnessed in those of a highly nervous temperament known as neurotics who have extremely irritable stomachs. To such it seems to act as a temporary sedative but in reality it produces a hyper-sensitive condition of the mucous membrane, which in time prevents the proper digestion of solid foods and has a tendency to add to an already over burdened nervous condition resultiug in enervation and prostration. There are also unknown conditions of ulcerations of the stomach where hot water often produces severe hemorrhages. These are only a few conditions in which the miscellaneous and indiscreet use of hot water has an evil effect. There are others too numerous to mention, but these will suffice to put the public on guard against foolishly and ignorantly aping a fad."

"If there be harm in drinking hot water under what conditions can it be used with benefit?"

A few instances where hot water may be successfully used are as follows: If upon awaking in the morning you find a sensation of fullness in the stomach, a heavily coated tongue, a slightly acid condition of the saliva you may know that your previous meal has left more or less of sour ferment in the stomach: Now, if you will drink half tea cup of hot water it will clear the mucous membrane of excess of acid, mucus and debrie remaining from the previous meal, making the stomach fresh and sweet for the morning meal. The principle upon which this acts is as follows: People who invariable eat three meals a day do not always completely empty their stomachs. Now the indigested portion of the previous meal remaining in the stomach undergoes a certain amount of fermentation, and if another meal be added without first clearing the stomach, the sour ferment remaing from the previous meal, has a tendency to decay the fresh food of the succeeding meal, thereby generating abnormal fermentation and gases which distend the stomach, with symptons of flushed countenance, slight palpitation of the heart and much discomfort. This is what is usually known as indigestion."

"Is there any way to make hot water more palatable?"

"Yes, it can be made much more agreeable to the taste to take boiling water and agitate it like making lemonade. If too disagreeable to the taste, a little milk may be added."

"What if it is not convenient to get hot water, doctor?"

"Then cold water may be drank a half hour before meals or especially at bed time."

"There is a popular notion abroad that hot water is a good drink at meal time?"

"Well, it's only relatively good, that is, it is not as bad as most drinks such as tea and coffee, but it has no especial merit to recommend it; on the contrary, any kind of fluid dilutes the digestive juices and makes digestion more difficult. The only so-called hot water drinks at meal time that may be said to have any merit is when as much milk is added as there is volume of water. Water makes the milk more easily digested and the merit may properly be said to be in the milk. Of course, if circumstances make it necessary to drink at meal times or not at all, hot water is the least objectionable, but it has no medicinal effect of importance, unless taken long enough before meals to allow it to escape from the stomach."

"About what temperature should hot water be drank?"

"Ordinarily from 105 to 110 degrees Fahrenheit, never hot enough to scald the membranes. The notion that boiling hot water is necessary is a grievous mistake."

CHAPTER V.

BREAD.

"Doctor, you say water in some form or other is the most indispensable of all foods, I suppose that meat is next in importance?"

"No doubt the Esquimeaux would say so, while those who style themselves vegetarians say that it is not only the least important but the most injurious of all foods, but the truth of the matter is, all races have lived off of what they could get the easiest and liked the best; but for the European and their American descendants, it can be truly said that bread is the staff of life, or more strictly speaking, the staff of life is wheat."

"I have often heard bread called the staff of life, but never knew why. I suppose it must be more wholesome or more nutritious than other foods?"

"That might be true in theory, but as a matter of fact it is often much more unwholesome than other foods. Aside from the fact that bread is both cheap and palatable it furnishes nearly all the essential ingredients to support life."

"Doctor, what are we to understand by essential ingredients?"

"First, heat or force producers—the starch and fat furnish these. Second, flesh formers or proteid food. This is furnished by the gluten of the flour. Third, mineral matter necessary to form bones and tissues. Fourth, waste material. Of course, bread is more or less deficient according to the material of which it is made."

"Some people think wheat the best, some rye and some Indian corn."

"Yes, the Russians and the Germans prefer rye or at least use rye, while most of the English speaking people prefer wheat bread, although in the Southern states corn bread is extensively used and preferred by many."

"The chemist out to be able to say which is the best, what food elements does each kind of bread contain?"

Fine flour ordinarily contains:

Force producers	Water	13.5
	Starch	73.2
	Cellulose	.75
Flesh formers	Fat	1.2
	Gluten	10.5
	Mineral matter	.85

It will be seen from this table that bread contains ordinarily about 7 or 8 times as much force producing food as that of tissue forming elements, a proportion considerably above what is usually estimated to properly nourish the human system."

"But, doctor. haven't you already said that rice was mainly all starch, and are there not more people who practically live on rice than an other article of food? If that be true, it doesn't seem reasonable to say that wheat bread really has too high a proportion of starch."

"It is true that more people live on rice than any other food, but an American laborer with his mixed diet can do twice the amount of labor in a given time than a laborer of rice eating nations."

"How about Mexico?"

"The people of Mexico eat meat and the Mexican laborer is in no way superior to the laborer of India or China, so that it is very difficult to draw conclusions by analogy, but this fact remains. The climate of Mexico makes a laborer lazy, sluggish and slow, and also has that tendency in India or China."

"Then on what does your statement rest when you say that ordinarily wheat bread has too high a proportion of starch?"

"It rests on a century of actual experience and it has been settled beyond dispute that a man requires a larger per cent of tissue forming food ordinarily called proteid or nitrogenous food than is contained in fine wheat flour, if health and physical development are desired, although the exact proportion depends upon climatic conditions, amount of exercise and the peculiarities of the individual. In Europe, the proportion of tissue forming food to that of heat or force producing food is estimated at a ratio of about 1 to 3. Some place the ratio as high as 1 to $4\frac{1}{2}$."

"How about our own country?"

"Well, Americans are the most active people in the world, and for the most part have rather a bracing climate, so that we can stand a diet as low in tissue formers as 1 to 6. Of course, this is speaking in a general way, extreme cold weather and active exercise might require even a higher ratio of heat producing food, while growing children in moderate or warm weather would require a proportion more nearly in accord with the estimates for the Europeans."

"What has climate and activity to do with heat or force producing food?"

"It has a great deal. It was at one time supposed that great activity destroyed a great deal of tissue, but that has been found to be a mistake. Hard labor or exercise increases circulation, and very naturally more heat producing food is oxidized, or burned up. The same reason holds good in cold weather. The need for heat increases respiration and circulation, and that burns up more fuel, which the heat producing food really is."

"Ah I see, this furnishes quite a guide to living. The sedentary and fat require less fat and starch than the active. In cold weather it requires more than in hot weather."

"Yes, that is the idea. The old soldier prefers a piece of fat bacon when he has a forced march, while to the aged and infirm it might be nauseous."

"I suppose we have about reached perfection in bread making, have we not?"

"I hardly think so; at least the masses are a long, long way from it, and there is probably no one article of food more responsible for indigestion, with its train of ills, than poorly made bread."

"I have heard some say that the best part of flour was bolted out—that our flour was too refined."

"There is a good deal of truth in that, for there are three important elements taken out—the bran, phosphates, mineral matter, and a considerable portion of the gluten and nearly all of the cellulose."

"Why are they taken out of the flour?"

"The bran is bolted out because it is unpalatable, and the phosphates are unavoidably taken out because they adhere to it. The gluten also adheres to the bran, but is mainly found in the heart of the grain—the part that grows. It does not pulverize so readily as the starch, and is also bolted out with the bran and known as middlings. There is another reason why middlings from a commercial standpoint is not desirable in flour—it makes the bread darker, but more yellowish than dark."

"Doctor, you haven't explained in what way the bran and phosphates increase the nutritive value of flour."

"I am coming to that. The bran has no food value, or rather nourishment for man. It is, in fact, indigestible cellulose."

"Then, I don't see any use in eating indigestible food."

"I'll tell you why.   Man no doubt originally ate much coarser food than he does now, and it is probable that his tendency is toward concentrated food; but, even if we grant that, it will still be several thousand years, if at all, before he can live on concentrated food exclusively."

"Then he needs indigestible cellulose in some form, as a sort of filling, for the same reason that a horse needs hay?"

"Exactly.   If there is not sufficient waste material, there is nothing to stimulate the action of the bowels, and constipation results, with all its attendant ills."

"Why wouldn't coarse vegetables answer as well?"

"But many people don't eat coarse vegetables, besides the waste matter can be too coarse. Many vegetables are stringy, and if hastily swallowed, which is a very com mon practice, they may really act as an obstruction rather than a stimulant to the bowels."

"But many people are troubled with diarrhœa rather than constipation."

"That's true; but most of those same people have constipation first, and the diarrhœa is only nature's way of getting rid of accumulated matter, and not a few persons have found that they lost their good health when their occasional diarrhœa ceased."

"There must be other causes for diarrhœa besides constipation."

"Yes.   They will be discussed under the proper head. I merely mentioned it to emphasize the value of bran for all people who have a tendency toward constipation."

"Is bran in bread of use to everybody?"

"Not by any means. If it greatly irritates the bowels, it should be avoided."

"Then, according to your statement, wheat bran is the best waste material found in any of our foods."

"I can hardly say that. The bran of other grains might be equally good. More depends upon its fineness than its name."

"What part of wheat is the richest?"

"That depends on what you mean by richest. A pound of wheat germs (wheat gluten) is more than equal to two and one-half pounds of lean beef, as flesh formers."

"Is the wheat gluten as easily digested as meat?"

"For a good many people, it is easier. Many persons tolerate it better than anything else, and it furnishes a good food at any period of life, and for almost any condition."

"As I understand it, the starch of wheat makes the fat; is a force producer; the gluten is the flesh former; the bran furnishes the waste material. Now, is that all the good things you can say about wheat?"

"No; the phosphates make the bones and furnish mineral matter for the system."

"Then, as a food, wheat seems absolutely perfect."

"It is said to be the only perfect food, and it probably is more nearly so than any other food; but for all that, it has too small a per cent of fat and too little lime. Pigeons fed on wheat and distilled water only lived a few weeks, but when water containing a small per cent of lime was furnished, instead of distilled water, there was apparently nothing lacking, and the pigeons grew fat."

"What would you suggest to add to it?"

"Well, I will talk of that when I discuss the different kinds of bread and diet suitable to meet different conditions."

"I infer from what you say that Graham flour or bread

made from it is much to be preferred to the ordinary fine flour bread."

"That is not the idea. Graham flour is supposed to be made of the whole grain, bran and all. Recently a process has been invented which saves all the valuable parts of the wheat without the objectionable and unpalatable bran in Graham flour. Wheat has three coats or envelopes, and it is advisable to remove the first two, which still leaves enough cellulose for a healthful diet, without being in the least unpalatable. Being sweeter, many people prefer bread made of entire wheat flour."

"Then, there is no general dislike to the new process common to brown or Graham bread."

"No; the dislike to brown bread results mainly from the unpleasant sensation produced in the mouth by the coarse bran, and if it could be reduced to the fineness of flour, there would be no objectionable taste. This has led to the late method of removing the coarsest part of the bran and the name 'entire wheat flour' substituted for Graham flour. It is not so white as bolted flour, but is really more palatable. Goodfellow gives the composition of fine flour and entire wheat flour as follows:

	Flour.	Entire Wheat Flour.
Water, ...	12	14
Proteids, ...	9-3	14.9
Carbo-hydrates, force producers, .	76.5	66.2
Fat, ...	0.8	1-6
Cellulose, ...	0-7	1-6
Mineral Matter, ...	0.7	1-7

"It will be seen, on comparison, that the entire wheat flour is richer in mineral matter, tissue forming elements and cellulose or waste matter."

"How would you sum up the advantages and disadvantages of entire wheat flour?"

"1st. Better for growing children, especially if there be constipation or tendency to rickets. 2d. The sedentary or corpulent. 3d. Vegetarians, or people who eat but little meat. 4th. People who suffer from constipation. 5th. Mothers during maternity, or while nursing children. 6th. Those who have a tendency to decay of teeth. No kind of bread should be given children under ten months old."

"Then, fine white bread is not so wholesome as the other?"

"That is not a fair way to put it. Much depends on individual peculiarities and what other food is used with it. Generally speaking, the entire wheat flour is much superior to fine flour. If exercise be such as to cause great peristaltic action of the bowels with looseness or diarrhœa, the fine flour bread is preferable."

"Doctor, you said a while ago that bread was a great source of indigestion. On what ground do you make the charge?"

"Well, bread may be very easily dissolved in the stomach, or may be very difficult. It depends upon its physical properties. If it be solid or sticky, it does not dissolve readily."

"Then, that would include pancakes."

"Yes, pancakes, dumplings, potpie, most pastry and all poorly baked bread. Any bread that will adhere together upon being pressed, forming a solid, doughy lump, is not easily digested and is a source of many disorders of the stomach."

"There are many arguments about which is more wholesome hot or cold bread. Which is right?"

"The wholesomeness of bread does not depend upon

whether it is hot or cold.   The objection to hot bread is
that as a rule it contains more moisture, and is therefore
much more doughy.   Its particles do not separate so
readily when put in the mouth.   For this reason there is
a tendency, almost universal, to swallow such bread in
sticky lumps, and of course the particles do not separate
easily when they reach the stomach.   This causes them
to be retained in the stomach so long that fermentation is
set up.   If bread not made with yeast is sufficiently well
baked, there can be no objection to it merely because it
is hot; but in yeast bread, unless very thoroughly baked,
the ferment does not leave the loaf until six or eight hours
after baking.   Biscuit should be thin and baked until its
particles will not stick together when mashed."

"What about cake?"

"Cake contains very wholesome ingredients, but made
well nigh indigestible by cooking.   Rich cakes might aptly
be described as butter, sugar and eggs, stuck together
with a little flour.   The general objection is that there is
an excessive amount of shortening which prevents the
digestion of the flour, and this is especially true if the
shortening be butter, because the amount of heat applied
in baking cake changes the chemical nature of butter, and
makes it very bad for people who have any form of dyspepsia.   There are still other objections: Heat coagulates any kind of albumen (by coagulation we mean condensing or hardening), and the time required for baking
cake necessarily so thoroughly toughens the egg it contains as to make it quite indigestible."

"What about the sugar in cake?"

"It may sour all that is eaten with it."

"Can you recommend doughnuts?"

"No; doughnuts are as indigestible as cake, for the same
reasons; but cookies are less objectionable than ordinary

cake, because they are not so rich; but fritters are probably the most indigestible of all cakes."

"Many kinds of light bread take their name from the
flour used and the methods of making light or spongy.
Yeast bread is most usually made of fine white flour, i. e..
flour made with bran and middlings bolted out."

"Doctor, from what you say, I conclude that flour or
wheat foods are all good."

"If not spoiled by the cook. It has been already mentioned that bread is often unfit to eat."

"Are there any reasons why bread is unsuitable for
food, other than what you have mentioned?"

"Yes, there are several faults common to ordinary
bread making."

"What are some of them?"

"Too much yeast is used, and too long fermentation
allowed. The more quickly bread can be fermented the
more wholesome it will be, and if fermentation be too
great, part of it is changed into acetic and lactic acid.
Bread is sometimes less wholesome because of ingredients
other than flour, which are added for various purposes.
Potatoes are often used, so that the bread will absorb a
large amount of water, making a heavy loaf with a small
amount of flour. Alum is frequently used in bread to
whiten it, and as it is an astringent mineral, likely to do
injury, no one should eat bread containing it. Another
extremely objectionable thing common to baker's bread
is the unwholesome places in which it is made. No language of condemnation can be too strong to apply to the
foul bakeries located in cellars and infested with rats,
roaches, flies, vermin, bad air from foul closets, and operated by an unclean baker. The health officers of
every city should see that all bakeries are kept in a sanitary condition."

"Can you give specific rules for bread making?"

' That is very difficult. Some flours require more kneading than others. Then, again, atmospheric conditions have something to do with it. Bread making requires care, and this is most likely the reason why so few bakers or cooks become good bread makers."

"A good many people say they cannot eat fresh bread only stale. Why is this?"

"The principal reason is that in fresh bread the particles adhere together in eating, so that it forms a large bolus, which is not easily dissolved by the gastric juices."

"Doctor, toast is nearly always used for the sick, and probably has been so used for several generations. Does toasting bread make it more digestible, or is it only used because it is more palatable?"

"It is both, but it is doubtful whether the reason is understood."

"Then, the fact that it is beneficial is only accidental, but the reason will be none the less interesting."

"It is very well known that dry charcoal will sweeten almost anything with which it is brought in contact—its disinfecting uses apply to the stomach the same as to other things. Now, toasting bread chars a certain amount of it, and it is therefore to that extent a disinfectant or sweetener, but the most important change is a chemical one, caused by the application of intense heat in toasting. Bread, when toasted, is changed to what the chemists call dextrine, which is part of the change that takes place in digestion, so that toasting bread partly digests it and makes it a more suitable food for those who are sick. It might also be added that the flavor of toasted bread is often very agreeable and useful on that account."

"Doctor, are there any objections to toasted bread?"

"There are no objections, other than the manner in

which it is done. To get good results the slices snould be cut thin and heat enough applied to drive all the moisture out of it. The heat should be applied slowly at first, and then finished at an intense heat. A good way is to put the slices of bread in an oven, and then only partly close the oven door. When bread is moderately dry it should be taken out and toasted. If the slices are cut thick and only toasted a little on the outside, the moisture in the bread is merely driven to the centre of the slice, making it much like dough and wholly unfit for the uses usually desired. For this reason toast, if not properly made, may be injurious instead of beneficial; and it should be borne in mind that bread should never be buttered before toasting, as butter melted by fire is chemically changed and injurious to the stomach.''

"How many different kinds of bread are there?''

"Probably the first and most common is ordinary yeast bread. The next in importance, or at least the next best known, is brown bread, or Graham bread. Recently the entire wheat bread is becoming much in favor. The various other breads are known as ærated bread, rye bread, milk bread, unleavened bread, malted bread, salt-rising bread, germ bread, gluten bread. This does not include every variety or patent bread, but substantially represents the different processes and kinds of bread. Others are mere variations of those enumerated. In yeast bread the sponge is formed by the yeast ferment setting free carbonic acid gas, which passes through the dough and makes it porous. Ærated bread is a chemical process. The gas necessary for making the dough spongy is generated in a separate vessel by the use of sulphuric acid on limestone, most usually marble dust. The gas is forced into water slightly acidulated, and this is mixed with the dough in strong receivers, where the gas is kept from escaping. After the mixing or kneading, the bread is baked the same as any other. This process made considerable headway for a time, but has been almost entirely abandoned of recent years. Graham bread is merely a mixture of common flour, bran and middlings. It is much coarser than ordinary bread, and is highly recommended in constipation. Its unpalatableness makes it unpopular. The entire wheat bread has already been described. Milk bread is made from ordinary dough, to which milk has been added. This improves the flavor and adds to its nutritive properties, but it does not keep sweet very long.

"Unleavened bread is not fermented at all, and no means are taken to ærate it by any chemical process. Flour is usually mixed to the proper consistency and then flattened out into thin cakes or strips, and baked quickly. Very nice biscuits can be made by mixing flour or milk at a very low temperature (ice cold), and baking in an unusually hot oven. The biscuits must be made thin, and the heat sufficient to quickly generate steam enough to make the biscuits almost as light as if chemicals had been used. Malted bread is made by the addition of barley malt to the sponge or dough. This quickens the fermentation and makes a very sweet bread; so that malted bread is said to be more digestible than ordinary bread, and the flavor more pleasant.

"Germ bread is made of that portion of the wheat known as the embryo, the part that grows. It is very rich in gluten, almost as much gluten as starch. It is therefore a specially valuable food for some people. Gluten bread is supposed to be pure gluten without any starch whatever. It is made by taking wheat middlings or flour and enclosing it in a bag or sack and washing until the starch is all dissolved and taken up by the water.

The gluten is not soluble in cold water, and therefore remains in the bag. The special value of gluten bread is because it contains no starch and is used in that class of disease known as glycosuria, or diabetes, and for obese people. It is also used in intestinal dyspepsia.''

"Doctor, you haven't lived in the South very much, or else you would have spoken of biscuits the first thing, instead of yeast bread. A good many people think them a great source of dyspepsia and unfit to eat, while there are others who think light bread unfit to eat, and refuse to eat anything except biscuits when they can be obtained. Will you tell which is better?''

"Biscuits are not objectionable because of their form or name, nor merely because they are eaten hot.''

"Then, I suppose you mean that they are objectionable only when they are not well made?''

"Not exactly. They may be very palatable, yet very unwholesome. There are two serious objections to biscuits—as a rule, they are insufficiently baked, so that they are sticky and become an insoluble mass when eaten; and the other objection is because of the chemicals used to make them light. Baking powders contain a great variety of ingredients, some of which are very unpalatable and nauseous. Many of them contain alum and ammonia. Alum is a dangerous astringent, while perhaps more real harm is done by ammonia. Whenever you break open a hot biscuit and can smell ammonia, it might be well to bear in mind that you have the equivalent of what you would most strongly smell upon entering a horse stable.''

"Aside from bread, the preparation of wheat most extensively used is crackers. There are many ways of making what is ordinarily termed crackers, but their food value is substantially the same, with a slight variation, according to the amount of shortening or sweetening that may be added."

"Some people think dry crackers very unwholesome?"

"In that they make a mistake. Being composed principally of starch, their dryness is an advantage rather than disadvantage, for they cannot be swallowed without mastication, and in doing this the particles of starch become more or less thoroughly saturated with saliva—as this is an important digestive agent for starches, it should be understood that it is better to eat starchy food dry. The United States army physicians say the soldier's health cannot be maintained without some dry substance on which to bite. The theory of this is that in using soft foods exclusively the uses and functions of saliva are partly or entirely dispensed with, throwing the digestion of starches entirely upon the pancreas and other solvents in the intestinal canal. This very fact of compelling one set of organs to do the work of another is one of the great sources of ill health. If people could only be brought to understand that starch is largely digested in the mouth, from the saliva received there, it might be possible to keep them from swallowing their food without chewing it. The ignorance on this subject is almost astonishing. Even a physician was recently heard to remark that he could not eat dry crackers; that his stomach was too weak to dissolve them. This no doubt was said without stopping to consider that starch is not digested at all in the stomach, only to the extent that the digestion continues for a short time from the effects of the saliva received in the mouth.

"Macaroni is an Italian preparation, and it forms a large part of their diet. It is made from wheat and generally supposed to contain a higher per cent of starch than flour, but recent analysis disproves this. Its stringy form is obtained by forcing dough through small perforations in a cylindrical sheet of metal. It is very similar to bread in its properties, and the only objection to it is that is not as friable as could be desired. By this we mean that the particles adhere together and do not separate readily by either cooking or by chewing, but if well masticated, macaroni is a wholesome, nutricious food."

"Doctor, are there not a good many preparations of wheat in the shape of meal, breakfast foods, etc.?"

"Yes, their use is just becoming understood, and too much cannot be said in favor of them. Even whole wheat, when it is washed and boiled for six or eight hours, makes a wholesome and nutritious food. Many people think it quite palatable when cream is added to it. Modern milling processes have given the people various preparations of wheat of the highest excellence."

"Which of the various wheat preparations do you prize most highly?"

"That would depend upon the use desired. Wheat germ meal (put up under various names of wheat germ, germea, breakfast foods, perhaps many other names) is one of the most valuable of all our foods."

"Doctor, you seem to give wheat the first place as a food. Are there any preparations of wheat you can specially recommend for different conditions?"

"There are a number of good wheat foods, and anyone who can increase their use as a food in place of many that are less desirable is a public benefactor. One of the concerns that has done this is the Purina mills of St. Louis."

"What is their product like?"

"It is a meal. They call it Ralston Health Club Breakfast Food, and though badly named, one can hardly say too much for it as a wholesome food. It is made from the very best wheat, the outer bran being removed by the cyclone process. This saves practically all the phosphates, which are usually lost in bolting out the bran by the old methods."

"Then, it is a whole wheat meal?"

"Not exactly. Part of the starch is removed in milling, making it richer in gluten. It contains less coarse bran than many other wheat preparations, and is therefore to be preferred for children. If hard water is used, the addition of cream makes it a perfect food. Such foods should not be used merely for breakfast, but should form a large part of our diet. Too much cannot be said in its favor."

"Doctor, what class of people are especially benefited by the breakfast foods, or wheat germ meal?"

"Perhaps its greatest use is for old people who are too corpulent; as the heat producing and fat forming element is less than bread, it furnishes the necessary elements of life without the objections to many other foods, meat in particular. It contains usually small particles of bran, which is a great aid in preventing constipation. It is also particularly valuable as a food for growing children, because it furnishes the necessary things to make bone and tissue. Its nutritive value, pound for pound, is about $2\frac{1}{2}$ times that of beef steak, and as it costs less than half as much per pound, its economy is apparent. It ought to a very great extent be substituted for meat, because it is a more wholesome food, and for many people even more palatable."

"How should it be cooked, Doctor?"

"It requires a great deal of cooking. Some preparations of it are partly cooked. These can be made ready for use in from half an hour to an hour. Like oatmeal, if it has never been cooked at all, it requires three or four hours continuous boiling to properly cook it. The same rules apply as that of oatmeal. It may be eaten with milk and sugar, but dyspeptics should eat it without sugar. Another preparation, not so favorably well known, is cracked wheat. This does not differ materially from the entire wheat kernel, but is more easily cooked, because it is partly pulverized."

"Doctor, the public will call you a crank on the value of wheat foods."

"Well, names don't hurt me,' while wheat foods help the people. So far, we have not discussed the cooked or predigested foods, and as some of these are so valuable as curative agents, they deserve more than ordinary mention."

"In what particular?"

"You will better understand their value when I explain that the modern way of treating disease is by aiding nature. This is done in two ways: (1) By increasing the activity of the excretory organs, and in that way throwing off the poisonous or waste matter from the system. (2) By furnishing the necessary elements for the body that will be readily assimilated, notwithstanding the enfeebled condition of the system. The Sanitarium Health Food Co., of Battle Creek, Mich., have made some new foods that better aid nature than anything heretofore known—at least, for some diseases, it is doubtful if there is any remedy equal to Granose."

"Granose! What is it? If it has such remarkable effects, the doctors will have to go out of business."

"There is another way of looking at it. If all died right soon, there would be nobody to get sick, but if they are kept alive they are likely to do some imprudent thing that will result in their illness, and Granose is so good for the sick that the people should know about it. As to what it is, I will explain in detail. Choice wheat is first cleaned of all dirt, chess, cockle and cut straws. It is then scoured and sterilized in such a way that all the starch cells are burst or broken apart. The next process is that of reducing the grains to thin flakes. This is done by machinery made especially for the purpose. The last important process is the roasting, which is done in such a way as to dextrinize the starch of the grain, and this makes a heat digested food.''

"That seems to be quite an innovation in the manufacture of foods.''

"Yes, the processes are quite original, each one of which are for a specific purpose. The object being to save all the valuable properties of the grain, convert what is commonly wasted into a valuable intestinal stimuant, harmless but effective, and at the same time make one of the most palatable and nourishing foods ever manufactured.''

"That is very interesting. What particular diseases are benefited by Granose?''

"Granose is the nearest a specific for constipation of anything yet discovered, and it will do more to smooth an irritable temper and clear away the clouds of despondency than any amount of good luck. People who have headaches, skin eruptions, asthma, epilepsy, piles, flatulent dyspepsia, torpid liver, or Bright's disease, should try Granose. Delicate and anæmic women and children will be greatly benefited by such foods, and if combined with the nut foods, rich and healthy blood will bring bright color and strength. If those who are subject to bilious attacks and sick headaches will eat such cereal foods as Granose, they will rarely, if ever, need drugs to keep them in condition. Granose is a good food in nearly all forms of disease, except diabetes and intestinal inflammations.''

"Doctor, you speak of Granose as almost a cure-all."

"Not at all, but so many diseases result from mal-nutrition it can safely be said that whatever relieves it must necessarily be a blessing to the race; and while Granose will not raise the dead, it is a valuable aid in restoring vigorous life in many forms of disease."

"Is there no other cereal food equal to Granose as an invalid food?"

"That depends on conditions. The same company make a food they call Granola, which is, as they style it, a twin of Granose. Granola is a mixture of wheat and oats, but prepared in a different way. It contains slightly more nutriment than Granose or other wheat foods, and is especially valuable for children, invalids, chronic dyspeptics, and those who have dilated stomachs or uric acid diseases. Both of these foods are ready for immediate use. A little water, milk or cream is all that is required to provide a health giving and appetizing dish."

"I don't see the advantage of having two foods so much alike."

"Different preparations, though equally good, supply different needs; besides there is a great difference in people's tastes, which is always an important consideration. To meet the great variety of tastes and needs of the afflicted, the Sanitarium Health Food Co. make a great variety of cereal foods. Crystal Wheat is another cooked food of much merit. It should be prepared in about the same way as rolled oats. Being coarser than Granose or Granola, it requires some cooking. It is a desirable food for daily use, and useful in dyspepsia and uric acid diseases. These foods are altogether unlike Graham flour, cracked wheat and oats, all of which have more or less sharp edge flakes of bran, which sometimes irritate delicate stomachs. The processes used in the manufacture of the

Sanitarium foods reduce the bran to such a fine state as to make them more palatable and less irritating."

"That all sounds very well, but the public will be slow to believe that any of these foods are nutritious like beefsteak."

"That is a subject on which the public are much misinformed. As a matter of fact, one pound of either Granose or Granola contains nearly three times as much nutriment as a pound of beefsteak."

"I don't see how that can be."

"Beefsteak contains so much water—nearly three-fourths—depending on how fat it is."

"From what you say, doctor, the making of health foods by the Sanitarium company is a great blessing to the afflicted,"

"Yes; in addition to the foods mentioned, they make numerous others that are extensively used. Their Gluten biscuit is of inestimable value for diabetics. They also make a 40 and 60 per cent Gluten biscuit, on which many diabetics thrive. So far as we know these are the only reliable Gluten preparations in this country. Their pure Gluten biscuit does not show any starch by the ordinary tests. As foods for the well, the Sanitarium biscuits (crackers) are both palatable and healthful. They make entire wheat and oatmeal biscuit with and without sweetning and shortening. Their plain biscuits, containing the entire grain with nothing else but a little salt are the best for dyspeptics and athletes. Their Gofio, Zwieback, Avenola, Wheat Granola are all useful cooked foods, while their Wheat Germ Grits is a good substitute for meats, and especially valuable to the corpulent and rheumatic with a tendency to constipation Of recent years many patent infant foods have appeared, and the Sanitarium company make a good one, but they do not recommend it for children who have not begun teething."

"Are there no other important ready prepared cereal foods?"

"Yes, the Sanitas Food Co., of Battle Creek, Mich., make a Malted Gluten that greatly aids in the cure of intestinal diseases, and as a food to tone up those who are run down. It is also of value in neurasthenia. Those who wish to get fat are sometimes greatly aided by first using such foods as Malted Gluten, or other food rich in nitrogen."

CHAPTER VI.

RYE AND CORN BREAD.

Rye bread has never been extensively used in America, although it is more or less used in cities having a German population. It does not differ greatly from wheat bread, except that it is darker, of a closer texture, and to most people less palatable. It is recommended to people who have a tendency to constipation. Doubtless this results more from its texture than its chemical elements, for various experiments have demonstrated the fact that it is less easily digested and more waste matter is thrown off. This perhaps explains its laxative tendency. It is said that it will keep fresh longer than wheat bread, and that it should be baked in a much hotter oven.

Corn bread is made from the meal of maize or Indian corn. It is used extensively in the Southern portion of the United States for bread. A large number of the people prefer it to wheaten bread.

"Doctor, a good many people think corn bread much more wholesome than wheat. Are they right or wrong?"

"There is some foundation for the belief, for as already explained, the tendency of English speaking people, at least, is toward foods entirely too concentrated, resulting in almost universal constipation, and the fact that corn meal contains more or less bran and is really a coarse food, explains why it is more wholesome than ordinary wheat bread."

"Has it any properties not common to wheat?"

"It has not. Corn is inferior to either wheat or oats as a food, except for fattening, although it contains very similar properties. It has a little more oil than wheat, and a higher per cent of starch, if grown in the Central or Western States. Corn grown in the far South has a high per cent of nitrogen or tissue forming elements. It answers the requirements substantially of either wheat, oats or rye, and is the cheapest of all foods, furnishing approximately the necessaries of life."

"What is the special value of corn?"

"Corn is an exceedingly nutritious food, containing all the necessary elements of food, but ordinarily too high percentage of starch, except when grown in hot climates. It is therefore the most fattening of all the cereals, or for that matter, it excels every other food in fattening qualities, except those containing large quantities of sugar or oil. Next to corn bread, corn meal mush is of secondary importance."

"Are there any objections to corn meal mush?"

"The principal objection is that it is too easily swallowed. Like all starchy foods, it requires the saliva to properly prepare it for digestion. Corn meal mush would, however, be much better than what it usually is if well cooked. It should be stirred in three or four times its volume of cold water, and then boiled for about three or four hours."

"What about corn starch?"

"Corn starch is made from both green and ripened corn. That made from the unripened corn is put up for the purposes of food. It is very palatable and nutritious, but contains no other element except the starch, and is therefore only a heat or fat producer. It is not a desirable food for persons who are troubled with acid dyspepsia, especially if it be eaten with sugar, as the combination • ferments quickly."

"What about roasting ears?"

"Green corn, commonly called roasting ears, is very

palatable and would not be objectionable except for the bran which envelopes the grain. This is a tough, insoluble substance, and is frequently a cause of diarrhœa or summer complaint during its season. If the starch be abstracted by grating, it is not so objectionable."

"Then, canned corn or dried corn would have the same indigestible material?"

"That is true; it is not good food. If we had some way of grinding up the husks or tough part, it might be especially valuable, but as it is, it is extremely objectionable."

"Are there no desirable preparations of corn?"

"Yes, grits and hominy are good foods, and are the cheapest of any in existence. Grits is the fine particles obtained in making hominy, and is extensively used for food in the Southern States. Hominy, being the coarser particles, requires if anything more cooking, and has proportionately a higher per cent of starch than the grits. Both should be boiled four hours, or until they are reduced to a pulp. Lye hominy is made from whole kernels of corn. It is placed in a vessel, and a weak solution of lye is added and left standing until the lye has in a measure destroyed the tough cellulose coat of the corn. It is then removed from the lye water, rubbed and washed until most of the bran is removed. Then the corn is soaked in water until the lye is substantially all absorbed."

"Does this make corn more digestible?"

"Yes; the lye has a chemical action upon the starch and partly digests it. Lye hominy is almost more of a medicine than a food, and it would therefore hardly be desirable for one in health to eat lye hominy continually but for some dyspetics it is especially useful."

"What can you say about parched corn?"

"Corn is parched by simply applying sufficient heat to roast it brown. The starch, in a measure, becomes dextrinized, and would be easily digested if it were reduced to a fine powder."

"Does the parching destroy the bran?"

"To a certain extent it does, and the only objection to parched corn is that it is usually poorly masticated, and if swallowed in coarse broken fragments, it is not easily digested. Pop corn is a small variety of corn used for food exclusively. It contains a larger per cent of oil than ordinary corn, and when subjected to a high degree of heat, the oil causes the grain to pop open."

"A great many people regard pop corn as unwholesome. If this be true, on what grounds?"

"Pop corn is not unwholesome so far as its composition is concerned, but its texture is where the difficulty lies. It is more or less tough and if swallowed in particles from the size of a grain of wheat up to whole grains of corn, it is very difficult to digest. It is therefore liable to cause either intestinal inflammation by irritation or obstruction in the bowels. It ought never to be given to children, and any one doing so cannot have much regard for the life of the child. It is not an uncommon thing among children for death to result from eating pop corn."

"Buckwheat is a cereal much prized by many people, but so little used it hardly deserves to be mentioned."

"What are its properties, Doctor?"

"The properties of buckwheat are very similar to that of rye. Both contain less tissue forming elements than wheat or oats, are less digestible, and not so valuable as food."

"I have never seen any bread made of buckwheat. Suppose it has no other use than for pancakes?"

"That is its common use. Some people like the flavor

of buckwheat cakes very much, but they are rather waxy and difficult to digest. There is another objection to buckwheat, and that is that it frequently causes an eruption of the skin called erythema. This is very disagreeable because of its continuous itching. No one should eat buckwheat cakes unless they care more for their palates than their stomachs and health."

"Cannot buckwheat be safely used at all?"

"It can be used with more advantage if nearly one-half the volume be of corn meal."

"Are these all the bread foods?"

"There are other vegetables used for bread, but not to any great extent in this country."

"What foods are suitable to be eaten with bread?"

"The relation of bread to our foods would include the whole system of dietaries, which should be formulated in another part of this work, although it is proper to remark in this connection that the cereals should form the principal part of our diet."

"Why do you say this?"

"Because they contain the necessary elements for supporting life, are less liable to produce disease, and are cheaper. Instead of making meat or common vegetables the principal part of our diet, the bulk of our food should consist of the cereals. If too low in fats and tissue forming foods, nuts, meats, milk or eggs should be added to sufficiently increase the nitrogenous and fattening elements."

"How much meat should one ordinarily eat per day, if the principal part of the diet be cereals?"

"The ordinary estimate for a person weighing 150 pounds requires about 8 ounces of lean beef to supply the necessary amount of tissue forming food. At least $\frac{3}{4}$ of of this amount would be furnished by a cereal diet, so that 2 ounces of lean meat will supply a fair average for amount of meat needed, exclusive of fat; but if entire wheat bread be used, with either milk, nuts or legumes, meat will not be necessary, except in disease."

"Doctor, what foods are incompatible with the cereals?"

"Acids of any kind are incompatible with starch. This is especially true of vinegar, and it is not advisable to eat rhubarb and fruits that are strongly acid at the same time that starches are eaten. They should be taken long enough before the meal or a half hour or so after the starches, but if no meat is eaten at all, no kind of acids should be taken at a meal composed mostly of the cereals."

"Does this same rule apply to sugar also?"

"Not as a general principal, but the fact that sugar is quickly converted into acid, makes it objectionable for those who have weak stomachs. It is almost certain to start an abnormal fermentation."

CHAPTER VII.

OATS.

"Doctor, are there any other cereals used for making bread?"

"Not to any extent. Buckwheat is used for making pancakes and oatmeal for crackers. In the tropical regions, there are fruits that are used for bread, but a discussion of these would be of no practical value. The cereal of the greatest food value next to wheat and corn, or perhaps second to wheat only, is oats."

"I notice that some people can't praise oatmeal too highly, while others condemn it. Perhaps you can clear up and explain these differences?"

"I think I can. Its food value has, so far as I know, never been overestimated."

"How is it then that there is such a conflict of opinion about it?"

"I can explain that. Oatmeal has a rough coat, and there is an occasional person whose membranes are so sensitive they cannot eat any food that has any irritating waste matter. The bran in any cereal either corn, wheat or oats, irritates the lining membranes of the digestive organs and causes such persons to have diarrhoea, although as a matter of fact, most diarrhoes come from constipation."

"Is there any was of overcoming the difficulty?"

"Only partially so. The oatmeal when thoroughly cooked, can be easily strained, though this will seldom be required except for young children, because most people need more bran or other waste substance than they get."

"But it is said that oatmeal is pasty and sticks to the stomach and is therefore hard to digest."

"It is rather difficult to get any substance whatever to stick to the stomach, so the trouble is not there."

"Where is it then?"

"There are two difficulties. The greater one is that oatmeal is seldom cooked half enough and most frequently not more than one tenth enough."

"Surely, you must be exaggerating, Doctor. The cooks of the country know more than that, don't they?"

"One would naturally think so, but they do not. The principles of cooking have never been given much, if any attention. It has always been merely an accidental routine or a striving to please the sense of taste. Most cooks merely bring food, water and fat in contact with heat, without much thought about results. It may be palatable but most likely indigestible. There is a more intelligent class who strive to make food palatable without much regard as to whether or not those who eat it will need the services of a doctor."

" But all you have said may be very true, but that doesn't explain why so much cooking is necessary."

"I have already stated that the cereals contain more starch than anything else. The starch is incased in tough cellulose sacks, or cells and they must be cooked enough to burst them. This requires considerable time."

"I suppose that the partly cooked preparation of oats could be cooked in a few minutes; at least, that's the general impression. If this is a mistake, I will venture to ask how much time is really necessary for proper cooking?"

"The so-called steel cut oats require at least three hours cooking, while the rolled oats should not receive less than one hour, considerably more will add to its digestibility."

"But some people say they don't like it that way. What say you to those?"

"That is largely a matter of habit. However, we ought not to be governed merely by what we like, for we might like what our reason would teach us was sure death, which we see exemplified every day."

"But it is urged that it gets too thick when cooked a long time."

"That can be obviated by adding plenty of water."

"What do you call plenty of water, a measure of oatmeal with an equal amount of water?"

"That's a common way of cooking it, and if it cooks too dry, which it is certain to do, they usually add more water; perhaps repeating the same operation several times, and then when it is served, you are very likely to say that you don't care for it, that it is not a palatable dish."

"What's wrong with the method just described and how much water does it take?"

"Oatmeal or rolled oats, is usually cooked in a double vessel, the inner vessel floating in water. The quantity of water depends upon the weight of the oatmeal, whether it be loose or packed, and also whether the vessel containing it is open or tightly covered. If the vessel in which the meal is cooked is covered so that no steam escapes, it will ordinarily be sufficient to add three measures of water to each measure of oatmeal. If the vessel be open and the atmosphere dry, considerable more than three times as much water as oatmeal should be used and always enough to cook it thoroughly without adding any additional water after it begins to cook."

"You haven't said anything about whether hot or cold water should be used."

"The cooks disagree on that point. If hot water be used, the oat flavor is stronger but much more care is required in making. If the oats is not sifted in the hot water very carefully, it is liable to be lumpy. It cooks more thoroughly in cold water, and this method is preferable from a standpoint of digestibility although there is no great difference."

"But why is cold water better than hot?"

"As already explained, the starch cells must be ruptured in cooking and they will absorb water more readily if cold water be used and the heat gradually applied."

"Is that the only reason for cooking it so long?"

"No. Intense heat applied to starch of any kind changes its chemical nature; in fact, it partly digests it."

"Doctor, I suppose that if oatmeal or rolled oats was prepared as you have directed, it would agree with everybody except those who are unable to eat any coarse food at all?"

"You are wrong. It may be perfectly cooked and still disagree—depends on how it is eaten and something on the pecularity of the individual."

"I don't understand you. Do you mean it should be eaten hot or cold, with or without sugar, at the beginning or the end of a meal?"

"Well, if proteid food (tissue formers) are eaten with starchy food, generally speaking, the starch should be taken first, so that the digestion of the starch will be as far advanced as possible before the stomach becomes acid, though this is unimportant. The greatest difficulty and objection to all soft, starchy foods is that they slip down too easily when taken in the mouth. I have already explained that saliva contains a digestive agent or solvent. This solvent is an important factor in the digestion of starch. Now, if the food is already moistened, there is no inclination to keep it in the mouth long enough for the saliva to be thoroughly mixed with it, and if it does not receive this digestive agent in the mouth, it is not digested until it passes through the stomach; most likely not at all."

"Wouldn't it be a good scheme to take a certain number of bites on each mouthful?"

"Yes, but rather hard to practice, but if you will try it, you will be surprised at yourself when you discover that you have been swallowing your food with so little mastication.   Bread should receive from forty to sixty bites or chews according to its texture, on each mouthful, and mush or porridge at least half as many.   Some people prefer to eat some solid food with their porridge, which of course increases the flow of saliva and makes it necessary to retain the porridge in the mouth for a greater length of time.   This partially overcomes the objection to soft foods."

"Have you ever thought of any other remedy for the constant tendency to eat too rapidly?"

"Perhaps where families are good enough natured not to quarrel, some system of small fines or forfeits for each one caught swallowing his food too quickly would work the best results."

"Doctor, you have criticised almost everything pertaining to the preparation of oatmeal, is there any other reason why it might disagree with people?"

"Yes, when it remains in stock too long it becomes vormy and when people are troubled with acid dyspepsia they should not eat sugar with oatmeal or for that matter, with anything else.   Oatmeal has about the right proportion of flesh forming elements to that of heat or force producing.   Now, if sugar be added, the proportion of heat or force producing element becomes far too high and whenever people live on food too rich in either tissue or heat producing elements the results will be disastrous. The fact that people do not know this is one of the prin-

‸ipal reasons why so many become ill. They do not have their food supply adjusted to their needs. The addition of milk and cream to oatmeal makes the food well nigh perfect. The cream supplying fat, in which oatmeal is deficient, and milk increases the proportion of tissue forming food and supplies some of the necessary mineral matter, which is particularly important for growing children."

"But, Doctor, most people think oatmeal unpalatable without sugar."

"That is also a habit. If they would eat it a few times without sugar, they would prefer it that way."

"But suppose they don't like it at all. You know there are many people who care very little for cereals of any kind."

"That is true, and is a matter which needs careful consideration. Much of the dislike as already indicated, results from improper cooking, and a dislike formed in this way is exceedingly hard to overcome. It sometimes happens that by the addition of fruit flavors, what would otherwise be unpalatable, is highly relished. Then, there are other ways. For people who like eggs, a very palatable dish can be made by stirring a raw egg into a dish of hot oatmeal."

"Doctor, I notice that some people say that oatmeal is a very rich food and that children should not eat it; that it is only suitable for those who exercise a good deal in the open air."

"It is rather difficult to see upon what such a statement could be based. The heat of the body must be kept up by some means, and starch, in which oatmeal abounds, is the least concentrated of heat producing foods. Sugar and fat both requiring much more air and exercise for their oxidation than starch; it follows that the charge that oatmeal is too rich cannot be sustained."

"How is oatmeal for old people, or those of sedentary habits?"

"Old people or persons who do not take much exercise, need food containing a larger proportion of tissue forming elements. This is especially true if they happen to be fat."

"Then the force producing food, is also fat forming and those people who are already too fat, don't need fattening food?"

"That is only true to a limited extent. No one can live for a long period without some fat forming food, but people who are very corpulent need much less than others."

"Is there any known reason for this?"

Yes. Layers of fat keep the heat in the body, so that a fat person needs much less heat forming food. Old people who are fat and sluggish need very little starchy food, but for active mechanics, farmers, laborers and growing children, oatmeal should form a considerable portion of their diet."

"Doctor, are there any other conditions in which oatmeal should not be eaten for food."

"Yes. In case of diahrroea or any inflamed condition of the bowels, no coarse food should be eaten."

"Doctor, it's a common practice of many people to eat fruit before breakfast, especially oranges, and then eat oatmeal, is this right?"

"No indeed! Oatmeal being principally starch, requires an alkaline medium for digestion. It ordinarily receives very little because it is swallowed so quickly and then if acid be added, there can be no digestion until it passes through the stomach, with a chance that it will not be digested at all."

'What foods are suitable to eat with oatmeal."

"No foods containing any considerable quantity of acid should be eaten for at least half an hour afterwards. This of course would exclude sour fruits, pickles or any dish on which vinegar is used. As to other foods, that would of coarse depend largely upon other conditions. One working in the timber with very severe labor, with temperature say 20 below zero, could well eat a great deal of fat and sugar as well as meat with oatmeal, while those who are corpulent and take but little exercise, would require food containing less starch or heat producers, and more tissue forming food; such as wheat gluten, peas, beans, milk, eggs, lean meat, oysters and cheese."

"Doctor, I suppose most of the barley that is consumed is taken in the form of lager beer."

"Not altogether so. Barley is used extensively for thickening soups."

"What are its properties?

Well, barley is mostly starch, perhaps 8 times as much starch as gluten. It is also rich in mineral matter. Before wheat became so universally cultivated, barley was a very important food, but now the only form known to the trade is that of pearl barley. It requires much cooking and in this respect it is very similiar to oats and wheat. There is still another use for barley and that is barley water. This is used extensively as a drink in cases of fevers, also useful for infants or invalids. It is made as follows: Grind half an ounce of pearl barley in a coffee mill, add 6 ounces of water, boil 30 minutes, add salt and strain. It should be made fresh daily and kept in a cool place. Another preparation of barley more used as a medicine than food, is malt."

"What is the process for making malt?"

"Malt is made by applying a considerable degree of moisture to the barley and allowing it to remain in a room

heated sufficiently warm to cause the grain to germinate.
It is then dried by different degrees of heat according to
the use for which it is intended.   During the germinating
period, a digestive agent known as diastase is formed.
This is both a medicine and a food and is used to great
advantage in diseases of the digestive organs where the
chief difficulty is the digestion of starch

CHAPTER VIII.

POTATOES.

"How did the Irish potato get its name?"

"I do not know. It was introduced into the Old World from the New by Sir Walter Raleigh and probably because it became so extensively cultivated in Ireland (forming a large part of the daily diet of the people) the name of Irish potato was given it."

"What are the properties of the potato?"

"The principal part of the potato is starch. It contains some waste material, and compared with other foods, a considerable amount of mineral matter, principally potash."

"How does the starch of a potato compare with that of other foods?"

"Very favorably. It is very similar to that of Indian corn but is not so fine as that of rice."

"Is there any advantage in the starch granules being very fine?"

"Yes, the finer the starch granules the more easily digested, although that might not always be an advantage. The tissue forming part of a potato is very small, exclusive of the water the proteid or tissue forming element is not much more than one twentieth of the solid matter. It will be readily seen from this, that the potato contains three or four times too high a ratio of heat producing food to that of the tissue formers. It is essentially a fat forming or heat producing food."

"Is this all it has to recommend it?"

"It is not. Potato has special uses. It has in addition

to the potash salts a small amount of citric acid. This is of but little importance of itself but the mineral matter altogether makes one of the best antiscorbutics known. By this is meant a food which counteracts certain diseases resulting from continual use of salted foods especially salt meats. The disease is seldom known outside of prisons and ships. At an earlier day, when voyages covered a period of several months scurvy was no uncommon disease on shipboard. It would seem only natural to associate the potato because of its potash salts with salted meats."

"Are there any other uses of the potato?"

"It is possible that the salts of the potato are useful in keeping the blood alkaline. Theoretically the potato ought to be very valuable in all genito-urinary inflammations, where it is desirable that the urinary secretions be kept alkaline."

"What about the digestibility of the potato?"

"If baked or boiled until mealy, it is quite digestible. If solid, or known as watery, the starch grains do not separate easily and is therefore rather indigestible."

"Then, this would indicate that fried potatoes are not wholesome?"

"If previously boiled and allowed to become cold and solid and then fried, as is usually done, they are not easily digested and not wholesome food, because being somewhat soft and waxy, they are swallowed in lumps and do not dissolve readily."

"Would not potato chips be still worse?"

"I hardly think so, being crisp they are much less likely to be swallowed without mastication. Besides, frying them brown, dextrinizes the starch and if ground up fine enough in mastication, potato chips should be fairly digestible."

"Don't the fat make them in a measure indigestible?"

"I am glad you asked that question, which would apply to many other foods but not to same extent to the potato, as both fats and starches are digested in the intestines and not in the stomach (further than what they are acted upon by the saliva), the fat would not therefore prevent digestion in the stomach as it would with fried meat or fried eggs. It would seem therefore that potatoes would be a good vehicle for the administration of fats. Usually fried potatoes are not sufficiently masticated and are a common cause of indigestion."

"Doctor, new potatoes are reputed to be the source of many digestive disturbances; is this true, and if so, why is it?"

"I suppose it is in a measure true. New potatoes are waxy and not easily dissolved. They might readily cause an irritation by remaining in the stomach too long, because they are in a degree insoluble. At any rate, new potatoes are not a desirable article of food, and it is a great deal safer for people in good health not to eat them at all or at least very sparingly."

"What is the best way to cook potatoes, Doctor?"

"The method to be preferred above all others, is baking. Boiling is also a very good method, but if cooked this way, it is better to boil them with their skins on than to peel them. They should be put in cold water and the temperature gradually increased. The third method is frying a potato in thin, crisp slices, known as potato chips, but as some people will not tolerate fat, frying would be objectionable to those."

"Then you are not a great enemy to the frying pan, Doctor?"

"I am very sorry to give anyone that impression, because the frying pan is one of the greatest enemies of the human race. Potatoes are about the only thing that is permissible to fry at all, and this is only allowable for people in good health who will thoroughly masticate them in eating.''

"Doctor, I perceive you do not rate the Irishman's friend as highly as some people. I apprehend that you will be severe on potato salad?"

"Well, I can't conceive of salad without vinegar, and vinegar and potatoes are about as incompatible as dogs and cats. Potatoes require an alkaline medium for digestion, while vinegar is a fermented acid."

"Doctor, you have said that potatoes are deficient in tissue forming substance, it would seem natural to connect them with meat?"

"Yes, the fact that potatoes are deficient in proteid and also in fat has led certain writers of large imagination to declare that the Irishman inseparably connects the pig and the potato, while the only necessary relation is the ease with which both can be raised. Many people get along very well who live principally on meat and potatoes, but eggs and butter or any other combination of fat and tissue food would probably do just as well to balance the defects of the potato as meat. People who live principally on potatoes have soft flesh and little endurance."

"Are there any other uses for the potato?"

"Starch is manufactured from it extensively, both for food and for laundry purposes. Various fancy names are given to potato starch for the purpose of selling it. It is very similar to starch preparations of corn and is equally wholesome and valuable for food."

"To what do you ascribe the universal popularity of the potato?"

"Its cheapness and the ease with which it is raised, together with the variety of ways in which it can be quickly cooked. These facts force its use until eating potatoes has become a fixed habit with the people, just for the same reason that where rice is easily raised, it is universally used as an indispensable food."

"Then you don't think much of the potato?"

"That conclusion is not warranted in anything I have said, because the potato is really a valuable food, but not equal to the cereals. It should, therefore, have a minor place in our dietaries and I can not urge an extended use of it."

"How does the sweet potato compare with the Irish potato?"

"Many people prefer the sweet potato. That is doubtless because it is sweet. Unlike the common potato it requires a warm climate and thrives best in tropical or semi-tropical countries."

"In what way does it differ from the common potato?"

"It contains less starch but a large per cent of sugar and gum. It is also more solid and stringy and requires much longer time to cook."

"I suppose you would call it a rich food?"

"Yes, it is both rich and heavy, for its particles do not separate so easily as most other starchy foods."

"What use has it as a food?"

"It certainly makes a very cheap food in warm climates. It is said that in South Florida they need not plant them but once. In digging up a row of sweet potatoes, they cover a portion of the vines between the rows and keep them growing perpetually in that way. Owing to the fact of its large percentage of sugar, as well as starch, it is a great heat producer and would be a food suitable for persons of good digestion doing hard physical labor."

"Doctor, I suppose that more people live on rice than any other article of food?"

"That is true. It is estimated that one third the people of the world live principally upon rice. In the United States, its use has never been so near universal as its merits deserve."

"What particular value has it?"

"It contains all the necessary elements for supporting life, but some in too small proportion. It is not so rich in tissue forming food as wheat or oats, and it is urged that because of this deficiency, the rice eating people are not so well developed physically as Europeans or Americans. It is also claimed that they do not so readily recover from an injury or a disease as those who live on a diet containing more of tissue forming elements. To offset this, rice is very easily digested, has the finest starch cells and is altogether a desirable food."

"Are the rice eating people more healthy than we?"

"They are at least free from some of the diseases due to excessive consumption of meat, because it is extremely difficult to overload the system on a rice diet, although one might become too corpulent."

"How does rice compare with potatoes?"

"Rice is far superior as an article of food for ordinary use to potatoes, although potatoes are much preferred in this country."

"From this, I conclude that the principal objection to rice is that people do not like it?"   •

"Yes, that is a serious difficulty. People like what they are brought up on, and the matter of eating different foods is largely one of habit. It is supposed that potatoes are much cheaper than rice, yet if we estimate potatoes at one cent per pound, and rice at six the difference would be very small. Potatoes contain about seventy-six per cent of water, so that one pound of rice is equal to about

four or five pounds of potatoes. and if the waste in peeling be deducted, considerably more would be required to
equal a pound of rice, so that people who have both to
buy, rice at some seasons would actually be the cheaper
of the two."

"How can the dislike for it be overcome?"

"That must be done by cooking and flavoring. Different people like different flavors and individual taste should
be considered. If nutmeg is agreeable, it may be added
so that it changes the taste of the rice and makes it palatable."

"How about rice pudding, doctor?"

"Well, instead of having rice pudding occasionally for
desert it would be better to frequently make it a considerable portion of the meal, and by varying the methods
of cooking and flavoring, the habit of eating rice could be
as well established as that of potatoes. This would avoid
the necessity for much of the meat or eggs ordinarily
consumed, and insure much greater freedom from disease."

CHAPTER IX.

PEAS, BEANS AND LENTILS, KNOWN AS LEGUMES.

"Doctor, of the foods discussed so far, the starchy element seems to predominate, and, with the exception of wheat and oats, the per cent of starch is much too high for perfect foods. Are there no vegetable foods containing a large per cent of tissue forming substances?"

"That question is answered by peas, beans and lentils. These are different varieties of the same species, or at least have very similar properties."

"Some have urged that beans should be substituted for potatoes. What do you think of the idea?"

"It is scarcely to be compared with the potato in any way, and has no such use, but it is used by vegetarians to a great extent as a substitute for meat, and to better understand them, we give the following table of analysis:

	Water	Protein	Fat	Heat Producers	Mineral	Waste
Butter Beans	14.84	23.66	1.63	49.25	3.15	
Peas (dried)	14.31	22.65	1.72	53.24	2.65	5.45
Lentils	12.51	24.81	1.85	54.78	2.47	4.58
String Beans	87.2	2.2	.4	9.4	.8	
Green Peas	78.1	4.4	1.7	16.	.6	

It will be seen that the variety known as butter beans, or Lima beans, has the largest per cent of tissue formers. That is, its relation to starch is greater than one to three. Peas and lentils are very similar in composition, only containing a little higher percentage of starch than butter beans."

"Is this for young peas and string beans or the dried?"

"Well, young peas have still higher per cent of tissue

forming food than the dried, that is, the starch develops in ripening more than the nitrogen."

"How do they compare with wheat gluten?"

"The per cent of starch is about the same proportion to the tissue forming substance in young peas as wheat germs, but it is not known as gluten but as vegetable casein."

"Which is the more easily digested?"

"The gluten is far more easily digested because not so tough and the particles are more easily separated."

"Then this must be much against them as an article of food?"

"Not necessarily so. While it is an objection for persons having weak stomachs, it may be of decided advantage to others."

"I don't understand that."

"Well, exercise of any kind has a tendency to develop strength, at the same time, what would be suitable labor for a person that was strong might easily cause the death of one that was weak. The same principle applies to the digestive organs. No food so difficult of digestion as peas and beans should be given to those persons who have weak stomachs."

"Then I suppose this is the reason why we should continually have in mind the digestibility of food."

"Yes, like exercise, it must be adapted to the strength or the ability of the individual; otherwise, our purposes would be defeated."

"Then where does the utility of foods difficult of digestion come in."

"A laborer of keen appetite and good digestion will be continually hungry if nothing but easily digested food be consumed, for such persons the legumes are particularly adapted."

"A great many people say that peas and beans cause flatulence."

"No doubt that is true. In addition to being tough and in a measure insoluble, there is a tough envelope covering the pea and the bean which is quite similar to the bran found in unbolted cornmeal."

"What effect has the tough envelope of the pea or bean on digestion."

"It is a very great factor in the disturbances common to their use, for it cannot be digested at all; consequently people who have weak stomachs will suffer more or less derangement because the pod or envelope retards digestion and prevents the food leaving the stomach as quickly as it should. This causes abnormal fermentation, and makes the gaseous discharges for which they are most unfavorably known."

"Doctor, is there any remedy for this?"

"In a measure, yes. Grinding to a fine flour is a great aid for it not only reduces the tough covering to a moderately fine particle, but it also separates the various ingredients and makes them much more digestible."

"Is this the only remedy?"

"No, there is another way of getting rid of the difficulty, and that is to boil six or eight hours and strain through a fine collander."

"What place should the legumes have in our dietaries?"

"It should have a very important place, especially among working people, also those who have fairly good health but occasionally have sick headaches, asthma, rheumatism and other ailments due to uric acid."

"Then you would substitute peas, beans and lentils to

a considerable extent for meat because they are more healthful?"

"Not that alone, although health is the first consideration. They are much cheaper than meat, although they do not furnish all the fat necessary for a perfect diet."

"Is it not true that starch makes fat? If so, why is it not a perfect substitute for fat, or oils?"

"Theoretically, that would seem to be true, but practically it is only a substitute to a limited extent, for it has been found that some fat is absolutely necessary to maintain good health."

"What foods are compatible with peas?"

"Well, as they contain a large amount of vegetable casein—tissue forming food—no lean meat of any kind such as beef, chicken, mutton, fish or even eggs should be eaten at the same time if any considerable part of the meal is made up of either peas or beans."

"What will be the result if they are?"

"An excess or nitrogenous of tissue forming food, which will be more than the stomach can properly digest and the system, especially the kidneys, will be burdened to throw off the excess. It should be borne in mind by all persons that an excess of tissue forming food is not so easily disposed of as either starches or fats, and those who habitually eat an excess of this class of food will quickly become what is known as bilious. There are many derangements of the system resulting from this condition that cannot be enumerated under this heading."

"What effect has acids on peas or beans?"

"Unlike starch, which is incompatible with acids, the digestion of peas, beans, or lentils, is aided by any of the ordinary acids."

"Then fruits may be eaten with them?"

"Yes, acid fruits, vegetables, cereals, butter, cream and bacon, but no lean meats."

"What would be a sufficient quantity of beans for one meal?"

"Well, a saucer full weighing 4 to 6 ounces, would furnish sufficient amount of tissue forming food for ordinary conditions without any meat or eggs."

"Doctor, you have forgotten to mention bean soup, have you not?"

"That is a good way to use the legumes. They should be boiled for several hours and then strained. Beans make a wholesome delicious soup in many respects, superior to meat soups, especially for those persons who are subject to uric acid diseases, such as sick headache, but soups should be avoided by persons of slow digestion or where there is dilatation of stomach."

"How about young string beans?"

"They are good if chopped fine enough so that the strings are not harmful; otherwise the tough fibrous threads interfere with digestion in the stomach and obstruct the intestines. For constipation there is nothing better among all the garden vegetables than string beans."

"Doctor what can you say about the properties of asparagus?"

"It is a vegetable that is used early in the season as the young tender shoots first put forth. It is quite similar to peas in flavor and is much prized by many people."

"In what way is it different from peas, Doctor?"

"It contains several properties not found in peas and has some medicinal properties. In composition, it is peculiar to itself, and while it contains a high proportion of tissue forming substance as compared with heat producing, it contains so large a per cent of water (over 93), it does not amount to a great deal as a food. In addition to the elements named, it contains gum, alittle sugar, resin, stringy fibers, asparagine, acetate, malate, phosphate and muriate of potash and lime, and nitrate of iron. It is diuretic, and is said to be somewhat irritating to the mucus membranes. Its effects and the reason for them are not very well known, but are supposed to result from the mineral substances set free during the process of digestion. It is usually eaten on toast."

"Doctor, what are the properties of Arrow Root and where does it come from?"

"Arrow Root is a starch extracted from a tuber that grows in the West Indies, principally Bermuda. The tubers are washed, dried, and then pulverized, and are nearly pure starch."

"Has it any particular uses?"

"Yes, it is easy to prepare, keeps longer than corn or potato starch, is a bland non-irritating substance and is much used for convalescents and in infant foods. Where there is inflamation of the stomach and the bowels are not affected, such preparations as Arrow Root and those of a kindred nature are of great value, because they give the stomach almost complete rest and yet furnish something on which to sustain life. It is not suitable for infants under 8 months of age; in fact, no solid food or starches of any kind should be given children until after they are eight or ten months old."

"What is tapioca?"

"Tapioca is very similar to arrow root. It is made from a plant known as manihot, which grows in Brazil, Central and other South American countries, also in the West Indies and Africa."

"Is it much used in these countries?"

"Yes, a flour and bread is made of it and it forms an important part of the natives' diet."

"What causes it's peculiar shape, Doctor?"

"The tapioca of commerce takes its form in consequence of the method of drying on hot plates. The heat used in drying bursts the starch globule thereby making them more easily dissolved."

"What food preparations are made of it?"

"It is used principally as a gruel and for pudding, and said to be particularly relished by infants at weaning. It does not sour as quickly as many other starches and is on the whole, a very pleasant and nutritious food, but requires, as in fact all starches do, some considerable tissue forming food or else the diet will be such as would cause disease of some kind. It is often flavored in various ways to add to its palatableness. Sago is another starchy food very similar to tapioca and arrow root. It is derived from the pith found in different varieties of palm in Java, Borneo and Sumatra."

"What special preparations are made of sago, Doctor?"

"Sago milk is prepared by soaking an ounce of sago in a pint of cold water for an hour or more and then draining off the water and adding one and a half ounces of milk. This is gradually heated until the sago flour is thoroughly incorporated with the milk. This adds much to the digestibility of the milk and makes a valuable food for persons recovering from fevers or other lingering illness. Sago gruel is made by soaking an ounce of the starch to each part of cold water for two hours then boiling for about 20 minutes. Sago does not differ much from either arrow root or tapioca, and the same methods of cooking and the same uses can be applied to all."

"What is Iceland Moss?"

"It is a lichen sometimes used as a food. It is made into bread in some countries. It's principal ingredients are gum and starch. It is also used in the food known as Blanc Mange. It has been recommended by some eminent physicians as suitable for use for diabetics. Many forms of sea weed and mosses are used by the Japanese and other people of Oriental countries, but very sparingly in this. Next to potatoes perhaps cabbage is the most favorably known of all the garden vegetables. It is used entirely in its green state. No method of preserving it except in sour kraut has so far as known ever been attempted."

"Of what use is cabbage, Doctor?"

"That is rather difficult to answer. It's a vegetable that is much relished by most people. It contains about 93 per cent water and would require a large bulk to amount to much as food. The properties are not in bad proportion, as the tissue forming elements being about 2 per cent are nearly one half the heat producing element, which is about 4 per cent. Cabbage contains the stringy fibers common to many vegetables, and this is the reason doubtless why it disagrees with many people. Boiling or cooking softens the starch and other elements of the cabbage, but does not destroy the stringy fibers."

"What harm do they do, Doctor?"

"Well, if they are cooked until they are soft, cabbage is likely to be swallowed in large stringy lumps as the fibers or strings are not easily dissolved, and it is very likely to remain in the stomach for a long time. If chopped fine and boiled without fat it is sometimes useful. People cook cabbage with fat meat and then eat it strings and all and I suppose that is the reason why so many people can taste the cabbage for half a day or a day after eating it."

"Is there any way of overcoming this difficulty, Doctor?"

"The only way it can be overcome is by grinding it up fine. It makes very little difference whether it be raw or cooked; if raw, it is usually tough and not likely to be re - duced to a very fine powder and if cooked, it is likely t) be swallowed strings and all. The only thing that can be especially said in favor of cabbage is that it is an antiscorbutic, which makes it especially useful in some classes of diseases. It contains considerable mineral matter, a large part of which is sulphur and if there is any malfermentation, the sulphur is the cause of the unpleas-ant odorous gases that are produced."

"What about sour kraut, Doctor?"

"Well, sour kraut is the most perfect representation of indigestible food that is known."

"What do you mean by that?"

"Well, as explained under the article on digestion, the digestibility of any food depends first upon its solubleness (how easily its particles can be separated) and second, upon how quickly it will ferment. Now, sour kraut is a tough, fibrous substance, and is extremely difficult to dis-solve, while the starch it contains is already in a ferment."

"Then according to this, Doctor, sour kraut is not a desirable food?"

"No, it is difficult to see where any conditions would arise that would require sour krout, and the best that could be said of it, is that it is an enemy of the human family, although persons doing hard physical labor and having strong digestion might eat some of it without injury.

"Cauliflower is of close kin to cabbage only more palatable, less stringy and altogether a more desirable food; when boiled, it is fairly digestible and people in good health may eat it, but those having a tendency to dyspepsia should not 'call for this flower.' As to Seakale well bleached, is about equal to cauliflower."

"The beet is a popular American vegetable. There are two varieties, the one known as ordinary beets and the other the sugar beet. The sugar beets have never been extensively cultivated in this country, except for a short period in Nebraska. In Europe the sugar beet is the principal source from which sugar is derived—a great deal of which is exported to this country. The ordinary beet contains about 90 or 95 per cent of water, a little sugar and a small amount of other matter, not important enough to take much account of as a food. It is therefore almost amusing to read the statement made by a scholarly man 'that young tender beets are very nourishing.' If by nourishing is meant that a half peck or peck of them would furnish enough sustenance to last a person for a day they may be considered nourishing."

"If there is so little nourishment in them what value have they as food?"

"A great many people are fond of beets. They make a very pleasant salad with vinegar and oil, and in that way may be used to a limited extent as a relish, but generally speaking, they have but little value as food and they have no other use worth mentioning except to fill the stomach when rich food is not desired.

"Carrots are classed with succulent roots. They contain between 85 and 90 per cent of water, 6 or 7 per cent of sugar a little nitrogen and a great deal of waste, which we ordinarily call stringy fiber. About the same can be said of the carrot as of cabbage. When they are cooked, one is likely to eat them strings and all. There is no objection particularly to them, if sufficient care be taken to guard against swallowing too much of the stringy fiber. The effect of this has already been explained."

"What can you say of parsnips?"

"The parsnip contains over 90 per cent of water, about 2 or 3 per cent of sugar and an equal quantity of starch. It has a rich flavor and a large amount of vegetable fiber. It is used extensively for stock food, but some people like the flavor very much and when young and proper precautions are taken to guard against eating the strings it contains, there are no particular objections to it."

"Then I suppose you would consider it a wholesome food?"

"Only for people who are in good health and who do hard labor. For persons of weak digestion it is likely to cause flatulence."

"I suppose you like the turnip because it has a "nip" at the end?"

"It sounds rather paradoxical to say that a 'turnup' should be turned down, although turnips are slightly more nutritious than carrots and parsnips. They contain about 85 per cent water, 3 per cent of nitrogen, 8 per cent of starch, and some mineral matter, nearly 2 per cent of woody fiber."

"Have they any uses for food?"

"Like parsnips and carrots, they might be useful for people who live largely on meat and need some coarse substance. It would be far better though, if we left turnips for cattle and depended more upon the cereals to supply the waste matter."

"What can you say of kohl-rabi?"

"It is an astringent vegetable and has no particular value as food.

Salsify, oyster plant, is a vegetable, which some people prize.   It is not extensively used but has some food value and is moderately digestible.   Artichokes is a tuber similar to that of the carrot.   It is usually known as Jerusalem artichoke.   It is said to be a much inferior in quality to many other tubers.   It is raised principally as a food for hogs, although it is occasionally used as a food for man."

CHAPTER X.

TOMATOES.

"Doctor, would you call the tomato a vegetable or a fruit?"

"The tomato is classed as a vegetable, although in reality it is a fruit. It is used much more extensively in this country than in any other."

"Is that because it is better known?"

"Hardly that, although the tomato has not been used for as many years as most of the other vegetables. Many people can remember when tomatoes first came into general use, and it is probably not more than 25 years since the modern varieties were introduced."

"What properties has the tomato?"

"Different analyses show different results, ranging from 89 to 96 per cent water. one to two per cent of tissue forming food, 1 per cent of mineral matter, and about 3 per cent sugar, starch and gum, with considerable wastematerials."

"Doctor, I was under the impression that the tomato had a great deal of acid in it?"

"So it has; the fruit contains malic and oxalic acids, the seeds oxalic acid, amounting to about half of one per cent."

"Has it any value as a food?"

"It could hardly be called a good food to use continually for several reasons. There is too much acid, besides the seeds are very objectionable. They are really sharp and have a tendency to irritate the mucous membranes."

"Is there any.hing that you can recommend them for, Doctor?"

"They certainly can be recommended for their flavor, for few vegetables equal them as an appetizer and under some conditions, they also aid in the digestion of proteid or tissue forming foods. This is because of their acid. They are also slightly laxative, due to the effect of the acids and seeds."

"I have heard people say that they cause cancer, is there anything in this charge?"

"That is nonsense, but it may have some foundation in this; that people do not distinguish between ulcer and cancer. The sharp acid that they contain together with the irritating seeds, might have a tendency to start an inflammation in which the seeds could collect and cause an ulcer, but an ulcer is so widely different from a cancer that they have no necessary relation.

"How should tomatoes be used?"

"Tomatoes may be eaten raw or cooked but they are really more of a medicine than a food. They should not be used at all when there is an acid condition of the stomach, and their use is very doubtful for any persons having rheumatism or a tendency to the formation of gall stones on account of the oxalic acid in them. It is better for a relish such as tomato catsup with the seeds removed than for any other purpose. They may be cooked and if so, should only be stewed in earthen or porcelain vessels, never coming in contact with any kind of metal. Canned tomatoes are sometimes bad on account of solder or poorly tinned cans leaving acid to come in contact with the iron, which makes a dangerous compound. No tin cans should be used but once, and it would be far better and safer if tomatoes were canned in glass or stone jars. They will keep just as well in glass as in tin if the jars are carefully wrapped with brown paper or kept in a dark place.

"Lettuce is a green vegetable of not much value for nourishment."

"What is its use then, Doctor?"

"Well, it contains a mild sedative substance which is useful under certain conditions and in some diseases."

"Will you please describe the uses and advantages lettuce may have?"

"Cases of diabetes have been reported cured with lettuce, although this is a matter in which mistakes might easily be made, but inasmuch as it does not contain any considerable starch or sugar, and is quite a sedative, it ought to be of great use in disease of the kidneys of the character of diabetes, although we would not be willing to stand on a declaration that anyone can be cured by a lettuce diet."

"Is there any other use for lettuce?"

"Yes, owing to its effect akin to that of opium it is said to be good food for sleeplessness and for that purpose should be eaten in some quantities late in the evening. It has a tendency to diminish action of the heart, and should be avoided where there is great danger of heart failure, although no attention ordinarily should be given to any danger so remote as this '

"Doctor, as celery is such a favorite with the people, no doubt you will will be able to say something very good about it?"

"Then you would have me treat the subject according to what the people believe and if I am only expected to tell them what they already know or what they think they know what is the use writing a book?"

"I confess that you have disarmed me, let us know the facts?"

"Celery contains some 90 to 93 per cent water, about

1 1–4 per cent of cellulose, 1 per cent of mineral matter, about 1 to 2 per cent of vegetable albumen and 4 per cent sugar and starch."

"According to that then it is very similar to cabbage?"

"Yes, but it contains less fiber and also less mineral matter, though it has a great deal of both."

"Then it is but little superior to cabbage except in flavor, but the people believe it to be a brain and nerve food, is there nothing in this?"

"Well, it is a pity to shatter their faith but there is no real foundation for the belief, except the bare possibility, that the mineral matter might be of benefit to those who are in the habit of living on food that contains but little of the mineral salts."

"Don't different foods nourish different parts of the body?"

"Not in the sense in which it is used. If they did brains would be brain food and we would all be wise. Food simply furnishes heat and material to replace the waste tissue according to the needs of the body, let that be wherever it may."

"Is there any objectionable properties in celery as a food?"

"Not more so than any other stringy food. It is simply a question of making it fine enough."

"How should it be prepared?"

"It should be chopped crosswise and very fine and then stewed until tender and served with milk, but the milk should not be allowed to boil, only gradually warmed for a few minutes?"

"What is the particular value of celery?"

"Its flavor. It is an excellent thing to flavor other kinds of foods less palatable. The seed as well as the stems, are also used for this purpose."

"May it not be eaten raw as well as cooked?"

"Certainly, only those who are in the least subject to ailments of digestion should take particular care to not swallow it with its strings. On the whole, celery is one of the most pleasant and appetizing of all garden vegetable, and as a relish deserves to stand ahead of any others, because it is free from acrid and irritating oil found in radishes, onions, peppers and other vegetables."

"Doctor, what are the various kinds of stuff that are used for greens?"

"Beet tops, onion tops, dandelion, sour dock and spinach. The latter is much more commonly used than any of the others. It is the only vegetable cultivated particularly for this purpose."

"What use have they as foods?"

"They have very little value as food, i. e. so far as any nutriment they contain, for they are principally fiber, and as ordinarily cooked and served with vinegar, they are exceedingly indigestible and likely to cause disturbance and irritation of the digestive organs."

"Why is this?"

"Well, because it would, as ordinarily cooked, be very much like eating a lot of hemp strings. If they are stewed until soft and simply swallowed, they will unavoidably obstruct not only the passage of food out of the stomach, but the intestines as well, and almost sure to cause flatulence."

"I have always heard that greens are very laxative and that they are frequently prescribed in constipation."

"Well, as they contain a large amount of waste matter, in fact very little else, they are naturally laxative, if properly used."

"How should they be served so as to overcome the objection you mention?"

"The leaves should be chopped crosswise until they are very fine and then stewed until tender, that would overcome the principal objection to them. Greens are of great value where there is torpidity of the liver and lower bowels. They are also useful in another way for diabetes, inasmuch as diabetics must live largely on animal food, it furnishes a coarse food substance to satisfy the appetite. Greens are also of value to people who are corpulent, for the same reason. Little nutriment, much bulk."

"What about serving them with vinegar?"

"Vinegar being a fermented liquid, it is objectionable to persons who have acid stomachs. If anything sour is required to make them palatable, lemon juice is much to be preferred."

"The onion is sometimes facetiously called the Irishman's fruit and whether this be a good name or not, it will not likely ever be called lover's favorite."

"Why so?"

"Because of the odor of a volatile oil which it contains."

"I never understood how the smell of the onion gave such an offensive odor to the breath after eating it."

"It happens this way; the oil or flavoring matter of the onion is taken into the blood in the process of digestion, then escapes from the blood because of its volatile tendency just as soon as it is brought in contact with the air in the lungs, and the breath coming from the lungs is laden with the smell of the onion."

"Is there any way to overcome the offensive odor?"

"Not entirely, though the fact that the oil is volatile a large portion of it escapes when the onions are cooked, as everyone knows that in cooking a kettle full of onions enough of the oil escapes to scent the atmosphere of the whole neighborhood."

"How should they be cooked?"

"Like most of the vegetables we have just described, it contains a large amount of fiber and is much benefited by being chopped crosswise so that the stringy substance is well separated with the knife. There is another advantage in chopping the onions in this way, and that is, they cook more quickly and mor e of the pungent oil is dissipated."

"How long does the smell of the onions remain in the system?"

"That depends upon the quantity eaten and the condition of the individual. If a considerable amount is consumed and the person is more or less constipated, the onion breath will continue for at least 24 hours; otherwise not so long."

"What about the nutriment of the onion?"

"The principal element in the onion is gum, with some starch and the average of several analyses shows about one part tissue forming substance to seven of heat producing, which is not a bad proportion. It also has some mineral matter."

"Why is it that onions disagree with people and that they can taste them so long after eating them?"

"Two reasons. One is, that if onions are eaten raw the oil acts as an irritant to persons of delicate stomachs, just for the same reason that radishes and peppers do. Then there is another reason and that is a raw onion is rather difficult to dissolve and considering its pungency and its toughness, it is no wonder that people taste them for some hours after eating them."

"Then according to this, Doctor, the onion is not to be eaten in the raw state."

"It is not desirable that way. When cooked, it takes a high place as a vegetable. It is somewhat stimulating to the system and supposed to be slightly laxative but not sufficiently so to deserve any special mention. Most people prefer cooked onions served with milk or cream."

"Garlick and leeks have similar properties to that of the onion. They have no especial value except as condiments to flavor other foods."

"The pumpkin is the largest of vegetable fruits cultivated in this country. It is said that the Indians cultivated the pumpkin with Indian Corn for centuries before the discovery of America."

"What property has the pumpkin that people should be called pumpkin headed?"

"That is probably because they are big and hollow, with nothing inside. I have often heard it said of them that they were all water, but this is not true, for about 2 per cent of the pumpkin is sugar. It contains a good deal of waste material, which together with the sugar makes it a valuable food, in connection with grain of some kind for fattening animals."

"What particular use has it as a food?"

"Many people stew pumpkin, but pumpkin pies are well nigh a universal favorite."

"Is there any objection to the use of the pumpkin as a food?"

"Not ordinarily, but for persons who have an acid stomach and need to avoid sugar and sweet foods generally, the pumpkin is not suitable, although it is a valuable food especially for growing children, who tolerate sweets and need something more or less laxative. The squash belongs to the same family as the pumpkin, although some varieties are very nearly identical with that of the sweet potato. These are baked and treated very much as a sweet potato. They furnish a rich and nutritious food but not particularly easy of digestion, and as they contain a very large per cent of sugar they would not be a suitable food where the pumpkin would not."

"The radish is a garden vegetable that has many staunch friends, for many people like it better than any other garden vegetable."

"What is the reason for this?"

"Because it contains a volatile and aromatic oil that gives to it pungency as well as flavor."

"Some people say that radishes do not agree with them, why is this?"

"The radish is sufficiently pungent to irritate the stomach; besides this, it is tough, solid and stringy. It is therefore difficult to dissolve and undesirable for persons of weak digestion."

"But, Doctor, there are other people who claim that radishes aid their digestion."

"That is rather doubtful, but if it be true, it could only arise from one fact, and that is that the eating of a small amount of the vegetable stimulates the secretion of gastric juice because it irritates the stomach while there is not sufficient amount of the tough insoluble part to seriously disturb digestion.

As a relish, to eat a bite or two, the radish may be of some use, and is certainly very pleasant, but no considerable quantity should be eaten by anyone, no matter how good their digestion may be. It is so near all water and fiber that it has no particular value as food."

"Pepper is a name for vegetables which includes quite a number of varieties, differing much in their degree of pungency. There is the common red pepper used only for sauces and then there are both sweet and pungent mangos. The mango pepper is used as a case for pickled cabbage. The flavor is much relished by many people, but it is exceedingly tough and indigestible. It has no value as food whatever and peppers do not deserve a place in any dietary as food, although they might occasionally be useful in a medicinal way for pepper tea."

"Rhubarb occupies a peculiar field among vegetables, for it has little similarity except in the manner of its growth."

"What are the properties of rhubarb?"

"Well, the rhubarb plant, including wine plant, is a stringy stalk containing a very large amount of acid and some gum."

"What kind of acid?"

"The acid of rhubarb is principally oxalic acid. When it is stewed a considerable part of it is dissipated. It is exceedingly stringy and objectionable on that account. Persons who have a tendency to an acid stomach should not eat rhubarb."

"Is this the only objection to it?"

"All persons who have a tendency to the formation of gall-stones or stone in the bladder should avoid rhubarb, because it may unite mineral substances in the system and greatly aggravate the tendency. It is not a desirable food but its acid may be useful when no other can be obtained."

"Vegetable marrow is a vegetable that is not extensively cultivated but one which some people like very much. It is so near all water that it is not especially valuable as a food. It contains a small per cent of starchy material, and a considerable amount of waste. Not much can be said either for or against it."

"Doctor, I suppose you cannot say much good of the cucumber, because few people do, except that they like it."

"That is true. A great many people prefer it to any other garden vegetable. Its flavor, like many other vegetables is because of the aromatic oil it contains."

"Why do cucumbers make so many people sick?"

"There are several reasons for this. It is possible that the aromatic oil which gives the cucumber its flavor, has some peculiar effect on the stomach, but this is probably not the main reason. It is difficult to say which deserves the most prominent mention, the toughness of the cucumber or the sharp seeds. More people are doubtless affected on account of the toughness of the vegetable, but when the seeds do cause mischief it is of a somewhat violent character. It is doubtful if there is any vegetable that is on the whole as insoluble as the cucumber, and the number of seeds is simply astonishing if they were all taken out, and what is peculiar about them is their sharp point and straw-like consistence. These sometimes prick the mucous membrane of the stomach and intestines but much more likely the latter. They are insoluble and liable to cause more or less obstruction in addition to the irritation. Of course, the cases in which the seed lacerate the membranes of the intestines are comparatively rare, but it deserves mention."

"What can you say about pickles?"

"Pickles are the enemy of the human race. If there be a personal Devil seeking the destruction and discomfort of the race, it is safe to conclude that one of his methods of operation is with the pickle."

"Why do you say that they are so bad?"

"Because of their insolubility and because they contain enough ferment in the shape of vinegar to disturb all persons who have weak stomachs."

"Is there nothing that can be said in favor of the pickle?"

"Absolutely nothing. There is something peculiar and abnormal about the craving for pickles and this is especially true of school children, who of all persons ought to be the last to eat them."

"How about the craving for something sour—for acids?"

"Well, acids should be furnished in fruits and not pickles."

"The watermelon has a very significant name, because it is more water than any thing else, and the season never gets so dry but what the watermelon is still composed mostly of water."

"Doctor, what is the composition of watermelon?"

"The part that is eaten is composed mostly of water, seeds, a little sugar and fiber."

"Why should the watermelon be objectionable?"

"Because it is often stale, tough and difficult to dissolve and is in that respect much like many of the vegetables. When it is well ripened, and has not stood too long after being removed from the vine, it is not more objectionable than sugar and water. It has about the same advantages with the addition of a small amount of flavoring matter which is supposed to act as a diuretic. The seeds of the watermelon are often used to make a tea for the same purpose and are useful in some diseases of the kidneys and bladder."

"The nutmeg or cantelope, is very similar to the watermelon, only it is richer. It contains even more sugar, is very palatable and a desirable food for those who tolerate sugar."

"Mushrooms are not used extensively in this country, because not much effort has been made to cultivate them and those found in the woods or field are difficult to gather, besides there is great danger of being poisoned by them. Some are even so poisonous that they will poison a person to handle them."

"How can the edible ones be distinguished from the poisonous ones?"

"They can only be distinguished by people who are familiar with them and have some knowledge of botany. There are three or four hundred varieties of edible mushrooms found in the United States, and the number of poisonous ones is also very large."

"What are the properties?"

"The mushroon is very rich in nitrogen, tissue forming substance—perhaps more so than any known vegetable."

"How is it then that it is supposed to cause indigestion?"

"Well, no doubt much more is charged to it than it deserves because it is usually eaten with other rich foods, but as it is usually fried this method of cooking would necessarily make it difficult to digest because the principle of frying tissue forming foods is radically wrong. Mushrooms, instead of being eaten with meat ought to supplant meat entirely whenever any considerable part of a meal is made of them."

CHAPTER XI.

SUGAR.

"Doctor, I suppose that most people know what is meant by sugar?"

"Yes, they understand that it is some substance extracted from plants and crystallized. They also know that it dissolves very readily and easily becomes a fluid."

"How is sugar obtained?"

"The sugar of commerce is obtained most largely from what is known as sugar cane, although much beet sugar is used in this country. The cane sugar comes from tropical countries while some of the beet sugar is produced in this country but more comes from Europe, principally Germany. There is also a small amount of sugar made from the sap of maple trees and from sorghum cane."

"Are there any other plants which contain sugar?"

"Yes, nearly all the plants used for food have more or less sugar in them and many fruits are very rich in sugar although the sugar of fruits is slightly different from that of cane sugar. Fruit sugar is known as levulose."

"Is there any sugar in meats?"

"No, but there is a great deal of sugar in milk, especially human milk. The milk in sugar is called milk sugar or lactose."

"To what class of foods does sugar belong?"

"Sugar belongs to the force producers and ranks next to fat and starch for that purpose. This being the case, it is aptly termed a concentrated food, for it has no waste."

"In what way is it useful as a food?"

"Well, the fact that it is found in most of our foods

would indicate that it had a high place in serving some need of the body."

"This will be very gratifying to persons who are said to have a sweet tooth, and they will be quoting you wherever candy is wanted and for every reprimand received."

"I am not so sure about that. It does not follow that because nature distributes sugar in most of her plants that the crystallized sugar of commerce deserves the same extensive use, and it is not clear why sugar should be added to our foods any more than it be would to add extract of beef to a steak or roast."

"I wasn't expecting such a turn as you have given the matter."

"So much injury comes from the excessive use of sugar that some one should warn the people of the fact."

"I never heard of the injury before."

"Perhaps you never thought about the needs of the system and how much sugar is really used."

"No, I never did."

"Well, considering the amount of sugar imported, the various kinds of sugar syrup and molasses made in this country, one is surprised how much per capita is really consumed."

"About how much will it make for each individual?"

"If we make some allowance for those who scarcely eat sugar at all and for small children, we are forced to conclude that the sugar eaters average from five ounces up to almost a pound per day. Now a pound and a half of sugar would furnish enough heat for ordinary exercise and weather, without any other food, and it will therefore be seen that a large amount of sugar in our diet dispenses with other foods. This leaves us a diet so deficient in many respects that injury could not help but follow."

"What is the effect of the excessive use of sugar?"

"Well, to understand that, we must consider that sugar is pure fuel with no waste matter in it and that every particle of it must be burned up in the system or thrown out of the system as sugar. Now, where much sugar is consumed it has a tendency to prevent the burning up of other foods, and this leads to serious consequences. If, however, the quantity of food is reduced in proportion to the amount of sugar used, then there is no waste material for the system, or at least it reduces the waste as well as the tissue forming foods. If other foods are not proportionately reduced according to the amount of sugar consumed, there may be conjestion of the liver, disturbance of digestion, excessive accumulation of fat, and finally diabetes. If the food be reduced according to the sugar, then there will likely be constipation."

"This is worse and more of it, is there any other mean thing you can say about sugar?"

"Well if sugar is not quickly disposed of, it will turn to vinegar, and in doing this it is likely to arrest digestion and if the food is not properly digested, decay sets in, which produces poisonous gases besides other decomposing substances that are likely to poison the system, causing languor, headache, rheumatism, and many other ailments.'

"I have often heard that sugar would cause the teeth to decay."

"There is not much in this, for the accumulation of starch and other food on the teeth would cause them to decay without any sugar, but a diet made up largely of it would be deficient in mineral matter and would therefore not furnish any natural supply for the bones, as well as other parts of the system."

"Is there any difference between sugar and molasses or syrup?"

"There is no difference except the quantity of water."

"How about candy?"

"Candy is flavored sugar with its crystals broken by
various methods of manipulation. Other substances are
often added, to increase the weight or change the texture.
These are not supposed to be harmful in themselves.
They consist mainly of such things as starch, and in very
cheap candy sometimes an earthy substance, chocolate, gum and other substances. Gum drops are less
likely to disturb digestion than candy."

•

CHAPTER XII.

VEGETABLE vs. ANIMAL FOODS.

"Doctor, is there any enlightened country in the world where the people eat as much meat as in the United States?"

"I don't think there is."

"Then if it be true that Americans eat more meat, and are, as it is claimed, more active and accomplish more in a given time, does'nt that prove the superiority of a meat diet?"

"No. It would be just as reasonable to attribute it to some one of a hundred other things. Esquimaux live on meat, and it would be illogical to say that meat eating made the people of the United States great and that it kept those of Greenland from any attainments whatever."

"The altruistic vegetarians are opposed to killing any animal for food, on the ground that man has no better right to live than the lower animals. What do you think of this doctrine?"

"It is a common thing in this country to call any one who is greatly interested in any subject, or who makes any innovation on existing things a crank, or a fanatic. This is wholly unwarranted, although it strikes me that the altruistic vegetarians practice extremely sentimental ethics. Life exists in every conceivable grade from the simplest vegetable to man, the highest animal. Who can tell just at what stage of development it is, or is not harmful to destroy it, although there seems to be general repugnance to each species destroying its own kind."

"Then you are strongly against vegetarianism?"

"If you merely mean their antagonism to meat, I am much in sympathy with them."

"I presume that you have some scientific reason for antagonizing meat."

"Yes, many of them, for meat, like alcohol, has important uses, but it is so much abused that it would be better for the race if its use were abandoned."

"But people rely on meat as the main source of strength and say they can not live without it."

"Suppose you tell that to the horse. There is no other animal that can stand so much or so varied physical exertion."

"That is so, but the digestion of a horse is better; that overthrows the point you make."

"I don't think so. It proves that the force or strength of the horse is developed and maintained solely on a vegetable diet, so that the charge that vegetables are not strengthening is here disproven, and if there is any fault it is in man's digestion or misuse of vegetables."

"But doctor, the people understand vegetables to be such foods as potatoes and cabbage, not bread."

"That is not a right understanding, for wheat and all cereals are vegetables."

"If it be true, as alleged by many, that vegetables are harder to digest, is not that a good reason for not using them exclusively?"

"Well, if we go on the theory that we should eat only easily digested foods, then the less effort required the better, and we ought, therefore, to eat nothing but predigested foods and thus relieve the digestive organs entirely."

"I don't quite understand you, for you have been continually denouncing indigestible foods and now you defend them."

"The point is this: as we exist in this age, our appetites are perverted and our digestive powers greatly weakened; these must be strengthened in natural ways."

"Then we are to eat foods adapted to our particular needs?"

"As far as possible. Take eggs for example. They are a good food, but it would be a perversion of nature to cook them with tobacco, and it is only less so to fry them hard in butter or lard. Now, one may require food harder to digest, and containing more waste than eggs, but it does not serve the purpose to merely make the eggs indigestible by some process of cooking."

"Doctor, your explanation is quite satisfactory and puts the subject of food in a different light from what I have ever seen it, but that does not explain what injury results from a meat diet."

"I can better explain the *use* and *then* the *abuse*. In discussing animal foods I endeavored to make it plain that owing to the chemical composition of meat it was not so readily burned up as other foods. Lean meat should not be used for heat production but only to supply the deficiency of nitrogen common to a diet of such vegetables as potatoes. For tissue forming food we rely on milk, peas, beans, gluten found in wheat, oats, rye and corn, especially Southern corn, and meat. Now two things govern the use of these foods; convenience of obtaining, and the idiosyncrasies of the individual."

"Then if milk, eggs or beans did not agree and gluten was not readily obtainable would you use meats?"

"Well, as a general rule gluten is preferable to meat, but not always. In continued fevers, like typhoid, meat powder and scraped meat are especially useful—the same may be said of chronic dysentery and some other diseases of the digestive organs."

"How about meat in diseases of the kidneys?"

"In diabetes, meat must be the main reliance for food, but in Bright's disease it is not permissible to use meat at all."

"Why is this?"

"In diabetes there is loss of sugar from imperfect oxidation of starches and sugars, while in Bright's disease the defect is just the opposite and the leakage is of albumen."

"What about the abuse of meat?"

"I have already explained that the waste of the tissue forming foods is eliminated by the kidneys. Now, ordinarily, ths system does not require that more than one sixth of the food be of tissue forming character, but a much larger per cent of such food is often eaten, and as most people stimulate their appetites with either condiments or liquors, it naturally follows that they eat too much."

"Then as I understand it, the excess of tissue forming foods overwork the kidneys?"

"Yes, excessive meat eating puts too great a burden on the kidneys, but this is not the worst effect."

"What is worse than disease of the kidneys?"

"The point I wanted to make was not the virulence of any disease but certain effects that are well nigh universal. If there be imperfect elimination of either the excess of meat eaten, or the dead tissue of the body, auto-infection will result with some of its numerous diseases."

"Why should meat be so much worse than other foods?"

"Because of its composition. You will better understand this by an illustration. Suppose you take five pounds of fresh beef in one vessel and five pounds of vegetables in another, then cook each and leave them exposed to the air in a warm room, what would result?"

"The meat would spoil in a short time and drive every one out of the house and almost out of the neighborhood."

"Then you don't think that the vegetables would greatly disturb any one when they spoiled?"

"No, and I guess I understand your point. You want to emphasize the fact that decaying meat is much more odorous than decaying vegetables."

"That is it. When meat or vegetables decay in the system, their relative effect is very similar to the comparative strength of their odors when decomposing, just as I have illustrated."

"Then you believe that a large per cent of the ordinary diseases are caused by excessive meat eating?"

"Undoubtedly; many people subject to bilious attacks, sick headache, rheumatism and other disorders have cured themselves by leaving off meat."

"Doctor, how do you get over the actual experience of laborers who say they can do more work on meat and even doctors themselves have tried the vegetable diet with unsatisfactory results."

"But more have tried it with satisfactory results."

"Then that would indicate that it was good for some and not for others."

"That may be a fact, and yet when we consider how little is known about the proper preparation of foods, and about their properties and uses, it is not surprising that a vegetable diet is not satisfactory. How many can tell the amount and properties of different foods, that would be required at different seasons for a perfect diet?"

"Certainly not many, for I interviewed more than one thousand physicians and only two of them could do it."

"Then how can they adjust a diet to their needs? Trying a vegetable diet, by bolting down, as is usually done, starchy vegetables (like bread and potatoes) into an acid stomach, would be absolutely certain to bring disagreeable results. Then there are other reasons for such failure. Many vegetables contain an excess of starch, and if fat and sugar be added as is usually done, the excess throws the diet too far out of balance."

"May there not be some peculiarity in the digestion of individuals, so that vegetables suit some best while meats are best for others?"

"That may be possible, and is what is called idiosyncrasy. People who feel discomfort from eating a large amount of meat or eggs will unquestionably do better on a vegetable diet. If the stomach only secretes a small amount of acid and pepsin, and the pancreatic digestion is good, vegetable foods will agree much better than meats."

"Don't some persons have strongly acid stomachs and weak pancreatic digestion, who can hardly be said to be diseased?"

"Yes, there are probably such persons, but the stomach specialist does not have occasion to treat such stomachs until there is disease, so that no one so far as I know, has sought such a subject on which to make a test. The starches in large quantities would not agree with such persons, but I am of the opinion that the vegetable gluten found in wheat, peas and beans if properly prepared would agree better than meat. The reason why people have failed with a vegetable diet is, that they have gone too far in substituting starch for meat instead of trying something like wheat gluten or nuts."

"Doctor, after all your predilections seem to be on the side of the vegetarians."

"No, I have no bias or fads, and I speak of things as I find them. My conclusions are based on many years careful study of the diseases of the digestive organs and how they are caused and influenced by diet and habits. A stomach specialist, who treats chronic diseases of long standing is compelled to study cause and effect."

"Then according to your view, Doctor, the main reason why the so-called trials of vegetarians failed is because those making the trials did not understand digestion, and the proper assimilation of foods?"

"Yes, that is it; one pound of wheat hearts contains as much of tissue forming food as two pounds of steak, and it is a fact that many invalids can digest it who cannot digest meat. and it is absur l to say that such foods are not sufficiently nutritious "

CHAPTER XIII.

ANIMAL FOODS.

"What is meant by animal foods?"

"Animal foods are not the foods of animals, as some might suppose, but they are the food products that are derived from animals."

"In what way are the animal foods different from the vegetable?"

"In many ways, although they contain some of the same elements."

"Which of the animal foods do you consider the best?"

"Milk is probably the first and most useful of all foods. At least, there has never yet been a satisfactory substitute as a food for infants, and as it seems to be Nature's method of feeding the young, we ought not, it seems, undertake to substitute a method of our own."

"Why is milk the best food for infants?"

"Because it provides all the necessary elements, not only for sustaining life, but for growth as well."

"Then, if it supplies all the needed elements, is it not equally valuable for grown people?"

"It does not follow that such is the case, because they do not require any material for growth, but they do require proportionately more of heat or force-producing foods, because the surface of the body being so much larger the radiation of heat is necessarily greater. In addition to this the exercise and labor of grown people necessitate an increase in food for heat or force production. An infant is kept within a warm room and does not require very much food for either heat or exercise. It needs mainly something on which to grow."

"What is the composition of milk?"

"The per cent of various matters in ordinary cow's milk is as follows:

Tissue-forming substance ..... 3 to 4 per cent.

Fat .... 4 per cent.

Milk sugar or lactose .... 3 to 5 per cent.

Mineral matter .... $\frac{2}{3}$ of 1 per cent.

Total solids being from ....12 to 15 per cent.

The remainder being water.

The analyses of human milk show a range of properties, as follows:

Tissue-forming substances .... 1 to $3\frac{1}{4}$ per cent.

Milk sugar .... $5\frac{3}{4}$ to 7 per cent.

Fat .... 2 to $4\frac{1}{3}$ per cent.

Mineral matter .... 1-5 to 1-6 of 1 per cent.

"It will be seen from this that a fair average would be:

Tissue-forming substances .... 2 per cent.

Milk sugar .... $3\frac{3}{4}$ per cent.

Fat .... 3 per cent.

Mineral matter .... 1-6 of 1 per cent.

In comparing these it will be noticed that cow's milk contains a much larger per cent of fat and tissue-forming food and a much smaller per cent of milk sugar, than human milk."

"Then, considering this fact, how can cow's milk be best adapted for infant feeding?"

"By adding about twice its volume of water, a little extra cream and enough cane sugar to increase the per cent of sugar in the milk to that of human milk."

"Why does milk disagree with so many people?"

"Milk disagrees with many people because it is not properly used; at least, that is the main reason. The protein or tissue-forming substance of the milk is known as casein, and the stomach has a milk-curdling element which at once coagulates the milk, i. e., separates the solid matter from the water, making it a solid mass of tough curds."

"What has this to do with its disagreeing with people?"

"Well, the particles of casein when formed in curds are too large for easy digestion by persons who have weak stomachs."

"Can this difficulty be overcome?"

"Yes; one way is to take milk in small quantities, swallowing it very slowly, mixing it with saliva, and, if it is necessary to use a great deal, it may be given quite frequently."

"Is this the only way of helping the difficulty?"

"No; there are many ways by which milk may be made easy to digest."

"What are they?"

"That depends upon the difficulty. Ordinarily the casein is the trouble, but it might be the fat, as some people do not tolerate fat very well. This is especially true of consumptives. If it is the fat that causes the trouble, it can be remedied by skimming. If it is the casein, which is most likely, there are many ways of preventing the formation of large curds in the stomach. One way is to dilute the milk with water. That is a very good way when it is not convenient to use any other. Another way, perhaps more important than all, is to dilute it with alkaline or aerated waters."

"What do you mean by alkaline waters?"

"Well, by alkaline water we ordinarily understand a solution of bi-carbonate of soda or lime water."

"How are these prepared?"

"Well, for lime water, take an ounce or so of slaked lime, about twice as much powdered sugar, and to these add a pint of pure water. The mixture should be shaken occasionally for some hours and then allowed to stand, and the clear liquid carefully drawn off."

"Which is the best, the lime water or the preparation of soda?"

"The lime water mixture is to be preferred when there is a tendency to diarrhoea."

"How is the water containing soda prepared?"

"Take ten grains of common bi-carbonate of soda (baking soda), about an equal quantity of salt, and a small amount of light magnesia. This may be added to a third of a tumblerful of hot water to two-thirds hot milk."

"What is the principle of using alkalies to dilute milk?"

"The alkalies neutralize the acidity of the stomach and prevent the quick formation of large curds and makes them more easily digested."

"Are there any other methods used for diluting milk?"

"Yes; milk may be diluted by barley water (see barley)."

"What is the principle upon which the barley water makes it easier to digest?"

"The fine particles of starch mix with the milk, and in that way separate the particles, or rather, keep them from forming too large solid lumps."

"I have often seen people boil milk. Does that add to its digestibility?"

"Boiling milk makes it more difficult to digest, because it coagulates the casein and renders it in a measure insoluble. It has this advantage, however: Boiling sterilizes it and arrests thereby, all fermentation, and by doing this the stomach may digest the milk, whereas, had it not been sterilized, the bacteria may have caused fermentation or decay. It is better to heat only to boiling point. Milk so treated will often be retained when all other methods fail."

"Does gelatine, or gum, make milk more digestible?"

"Well, some recommend mixing gelatine or some of the gums like gum arabic. If white gelatine, such as the Keystone, be soaked until dissolved and then sufficient water added to make it pour readily, it makes an admirable milk diluent."

"The gelatine should be soaked in cold water for several hours and then the cup set in water and boiled; then it is fit to add to the milk. A teaspoonful may be put in a nursing bottle with two or three ounces of milk."

"How is the gelatine prepared?"

"Why does taking acids and milk together frequently make people sick and cause them to vomit?"

"Almost any kind of acid, whether it be fruit acid, vinegar or the mineral acids, will coagulate milk. That is very noticeable if it be used on cherries or sour berries. Now, if a considerable quantity of acid be taken with milk, it forms large clots or curds in the stomach, and if the stomach happens to be a little sore, the clots irritate it sufficiently to cause vomiting."

"Then, according to this, no one should ever eat acids and milk together?"

"No; they are wholly incompatible."

"What about cream?"

"Cream is that part of the milk which rises to the top of the can after it has stood for some hours. The reason it does this is because the fat or oil is not as heavy as the milk."

"What is the ordinary composition of cream?"

"Well, cream ordinarily contains about two-thirds of its bulk of fat, one or two per cent of casein or tissue-forming food, two or three per cent milk sugar, and a trace of mineral matter."

"What are some of the particular uses of cream?"

"Cream has many and varied uses. There is no fat or oil which ordinarily agrees with a disabled stomach so well as cream, although in some constitutional maladies cod liver oil has greater value. The reason for this is, that cream is one of the most easily digested of all the fats because its particles are more readily separated. It is of great use in diseases of the stomach where digestion can be performed in the intestines, and as fat is not greatly acted upon by the secretions of the stomach, cream gives the stomach rest, and furnishes a large amount of heat for the body. For fattening purposes, cream is especially desirable, and people who wish to put on fat for their comfort or their beauty, can often do so more quickly by using a large amount of sweet cream, than in any other way. Pure cream is not affected by acids to the same extent as milk, but milk and cream as ordinarily used is incompatible with acids."

"Doctor, I have heard a great deal about milk infection; is it really a serious matter, and if so, what are some of the causes?"

"Yes, milk infection is much more serious than the people suppose, because there is no food which so readily takes up poisonous bacteria as milk."

"How does the infection ordinarily occur?"

"One of the great sources of infection is from the vessels in which it is kept or handled. Typhoid fever has been spread in many cases by washing the cans from a well that was infected, and while the cans would appear to be perfectly sweet, they still contained deadly poison. Then again, milk will absorb poison in cellars containing foul air or in a sick room, where there is disease.

Another source of contagion is from the animals themselves. They are often kept in filthy, disease-breeding stables, milked by persons whose hands are perhaps both diseased and filthy, and then the milk is often allowed to stand around in open cans and buckets, in foul-smelling stables and yards."

"What is the remedy for this?"

"Of course the greatest remedy would be cleanliness, but as the people who usually supply milk are beyond the reach of those who buy it, the only thing to be done is to strain it and treat it in such a way that disease-breeding bacteria will be destroyed."

"How may this be done?"

"Doubtless the best way to strain milk and be sure that it contains no part of barn-yard filth, is to take a piece of cotton, sterilize it (by boiling) and then put it in a funnel and strain the milk through it. There is also a process of purifying by centrifugal force. The most popular way for rendering inert any germs in milk is by pasteurizing, though it is alleged that infants fed on pasteurized milk · have developed rickets."

"How is milk pasteurized?"

"Well, in substance it is keeping milk at a temperature of about 160 to 170 degrees Fahrenheit for a half hour or more and then allowing it to cool."

"Does this greatly affect the character of the milk?"

"No. The change in taste is very slight, probably not noticeable at all. For ordinary use the best way to sterilize milk is to take bottles that have been cleaned with boiling water; then take the corks and clean them thoroughly with boiling water and punch a small hole through them. Fill the bottles with milk and then take a kettle of boiling water and add a small amount of cold water to reduce the temperature slightly, and set on top of a stove or where it will get only limited amount of heat. Put the bottles of milk up to the cork in this kettle of hot water and allow them to stand thirty or forty minutes, and then stop the hole in the cork with hot wax or sterilized cotton. The water should not be allowed to get cooler than 175 degrees F."

"How long will milk keep if treated this way?"

"Pasteurized milk has been kept sweet for a year or more, but the ordinary precautions taken are not sufficient to insure the absolute destruction of all bacteria, but if it be done with any care at all, the milk will keep several days, if put in a cool place."

"What is the difference between pasteurizing and sterilizing?"

"Well, in pasteurizing the average temperature is about 165 degrees F. This temperature if kept up for some length of time destroys ordinary bacteria, but not all germs of every description. To sterilize milk it must be raised to a temperature of 212 F., which changes the taste very much."

"This is merely a process of boiling?"

"Yes; the reason why people do not succeed better with these processes is from the fact that after sterilizing the milk they put it in vessels that are not sterile; whereas they must not have been only sterilized, but the cover also must be sterile. In addition to this they must be sealed up air-tight as soon as they are taken out of the sterilizing apparatus. It is not much use to sterilize milk and then pour it out and let it stand in an open vessel."

"What uses has milk as an article of diet?"

"Its principal uses will be discussed under various diseases, but milk has great value as an article of food in health, as well as disease. It is not, however, suitable for an exclusive diet for grown people, because it does not furnish a sufficient quantity of heat-producing material, but being rich in tissue-forming substances there is no food equal to milk for growing children; it also furnishes

the most available and one of the most desirable additions
to either cereals or vegetables."

"But most people say it makes them bilious?"

"There is some foundation for this charge from the
fact that milk does not contain sufficient waste material
for most people, and as it is usually drunk in bulk, a glass
at a time, it is apt to form large curds in the stomach,
whereas if it is only sipped, a spoonful or swallow at a
time, the difficulty would be entirely overcome, and if
treated by any of the methods described, there are very
few people with whom milk cannot be made to agree."

"How can its use be varied so that people will not get
tired of it?"

"There is no particular objection to flavoring it with
anything that may be agreeable, such as nutmeg or cinnamon, and if necessary to use it extensively, it may and
should form the principal part of puddings."

"Then you do not consider milk of itself a good drink?"

"It is not a good drink for many people, although if
equal parts of hot water and milk be mixed, there is no
other drink that can be taken at meal time that can be
compared with it. The merit, of course, is in the milk,
and the hot water merely prevents the formation of large
curds."

"What other kinds of milk are used in addition to cow's
milk?"

"Goat's milk, jennet's milk and mare's milk. Koumiss
was originally prepared from the latter in Russia, but its
use has extended over all Europe and America, and it is
now made of cow's milk."

"Are there not many kinds of foods derived from milk?"

"Yes; butter is the most favorably known. Some one
has remarked that if bread be the staff of life, butter is
its golden head."

"Does butter agree with people better than other kinds of fat?"

"Butter is to be preferred to any other fat in general use except cream, although a very great amount of digestive disturbance results from the improper use of butter."

"How is that?"

"Well, if butter be poorly worked a considerable amount of milk remains in it which soon becomes rancid. This is of course easy to detect and the people ordinarily refuse to eat it. But while this is true, it is a common notion that rancid butter is all right for cooking purposes, and it is no uncommon thing to hear people inquire of produce dealers for cooking butter."

"What wrong is there in this?"

"No wrong, I suppose, in inquiring for it. The wrong is in using it. In the first place, butter ought not to be used for cooking at all."

"Why?"

"Because heat bursts its fat globules and sets free both lactic and butyric acid, and if any one wants to be certainly dyspeptic all they need to do is to eat plenty of fried food cooked in rancid butter, or, for that matter, butter of any kind.

"How about buttermilk? Some people say that buttermilk is more wholesome than sweet milk."

"It contains about the same properties found in sweet milk, except, of course, that it is partly fermented and has lactic and acetic acid."

"Is not this a serious objection?"

"That would depend upon the kind of bacteria that caused the fermentation."

"Why, then, should sour milk be more easily digested than sweet?"

"Only for this reason: that in becoming sour, the particles of casein are much more evenly distributed than if taken into the stomach and curdled there by the acid of the stomach; so that sour milk has the curd broken and might be more easily digested on that account."

"Is not sour milk also used for making biscuits?"

"It is; but it is always more or less difficult to know just how sour it is, so that too much or too little soda may be used."

"Of what is cheese made?"

"Cheese is made of milk with or without cream. The milk is curdled by rennet, which is taken from the stomachs of slaughtered animals."

"Doctor, cheese is said to be a very rich food."

"So it is, for it contains some thirty to forty per cent of tissue-forming substance and from seven to thirty-five per cent of fat. It also contains a small amount of milk sugar and ordinarily about four per cent of mineral matter."

"Why is is that cheese disagrees with so many people?"

"It has been well said:

'That cheese is a bewitching little elf,
Digesting everything except itself;'

and I have siphoned out the contents of stomachs six hours after meals and found everything digested except the cheese."

"How do you account for this?"

"Well, cheese is a decayed food and probably excites a greater flow of gastric juice than any other common food. The reason it is not easily digested is because it is thoroughly infested with bacteria that have passed through an acid fermentation, and are, therefore, not readily acted upon by the gastric secretions."

"Is cheese, like milk, incompatible with acids?"

"Yes, more so; this is especially true of tannic acids

found in blackberries and raspberries.   Cheese eaten with these will often cause an attack of catarrh of the stomach. Very strong tea is incompatible with both cheese and milk on account of its tannic acid."

"Doctor, it looks as though you had cut cheese out of all dietaries."

"So I have; it is not suitable to eat with starchy vegetables; but a small quantity might be eaten with meat, beans or peas."

"Then crackers and cheese don't go well together?"

"No.   If a meat and egg diet be necessary, cheese might be occasionally added to an advantage; but it does not deserve an important place as a food, and must not be used at all when the stomach is inflamed."

"Pot cheese contains about twenty-five per cent of tissue-forming elements, about seven per cent of fat and considerable mineral matter.   When freshly made, it is said to be very wholesome and digestible, and quite similar to buttermilk."

"How is condensed milk made?"

"Condensed milk is made by evaporating ordinary milk at a low temperature until it is about the consistency of honey.   It has all the properties of milk and usually has about forty per cent of sugar added to it."

"What is the use of adding the sugar?"

"The sugar preserves it and makes it keep better than it otherwise would."

"How long will condensed milk keep?"

"Properly sealed, it will keep almost indefinitely."

"What is the use of condensed milk?"

"It is used extensively on voyages and under other circumstances when it is not convenient to get fresh milk. It should not be substituted for fresh milk except when unavoidable, but it is preferable when good milk is not obtainable, or cannot be kept sweet."

"Malted milk is made by evaporting milk similar to the method of condensing it and then adding the malt, which is a digestive agent made from barley and wheat. Malted milk has great value in many diseases."

"Evaporated cream is the same as evaporated milk, except that only half or two-thirds of the cream is removed, whereas, in ordinary evaporated milk, it is all removed."

"There is much discussion about the healthfulness of ice cream."

"That can easily be, because it is used to considerable extent in diseases of the stomach, especially ulcer; at the same time, it is also the source of many digestive disturbances."

"How do you harmonize these conflicting effects?"

"Well, it is this way: Ice cream is made of wholesome and nutritious ingredients and where there is inflammation, and the stomach in a condition that no solid food can be taken, it has a soothing effect; but ordinarily, ice cream is eaten with much other food and entirely too quickly, because pleasant to the taste, and easily swallowed. The stomach was never intended for a refrigerator, and when so used, it is often very disastrous, because it arrests digestion, and to a certain extent paralyzes the nerves of the stomach, causing languor and headaches, and very often catarrh of the stomach and diarrhoea."

"Then you would strongly condemn it, or at least as ordinarily used?"

"Well, certainly as it is now used, it does much more harm than good; but if eaten very slowly on an empty stomach, or with very little other easily digested food, like dry crackers, there is no reason why ice cream should seriously injure any one, but to people who will swallow a spoonful at a time and eat two or three dishes, there is almost certain to be ill effects following its use."

"What is milk shake?"

"Milk shake is made from ordinary milk, to which various flavoring substances have been added to suit the taste. It is then agitated at a very rapid rate, usually with a machine constructed for the purpose, until it is thoroughly aerated."

"Is it a good and healthful drink?"

"It is; the aeration adds much to the digestibility of the milk. It is very palatable, wholesome and nutritious."

CHAPTER XIV.

MEAT.

The propriety or impropriety of slaughtering animals for food, according to the altruistic views of vegetarians, does not come within the proper scope of this book; hence, we have nothing to say on this subject."

"Then you believe in treating foods according to their merits?"

"Yes."

"What place does meat deserve among our foods?"

"Speaking in a general way, it would not be far wrong to say that it deserves the long side of neglect."

"Why so?"

"Because its use is so much abused."

"In what way?"

"By cooking it until it is indigestible and then eating from three to six times as much as the body needs."

"Well, Doctor, if you can demonstrate that, the butchers and doctors will both be after you for damage done their business."

"I don't see it in that way, for the people will have just as much money to spend and the doctor and butcher will get just as much of it as they do now, but in some other way."

"But where is your proof?"

"Well, to begin with, let us examine the composition of beef. It ranges as follows: Proteid or tissue-forming substance, from thirteen to twenty per cent; fat, from ten to thirty-three per cent; mineral matter, from one to three per cent; water, sixty to seventy-five per cent. This estimate is made without the bone. It will be seen from this, that the two principal elements of beef are fat and tissue-forming food."

"Is the fat valuable?"

"Not especially so. Beef fat is much more solid than many other kinds. It is not particularly pleasant to the taste and has nothing to recommend it. Cream and butter, and many other fats, are better for general use."

"Then the value of meat must be in its tissue-forming element?"

"It is principally so."

"Then what is your objection to meat?"

"Well, as already explained, foods only serve two purposes; that of repairing the waste of the body, and furnishing it with fuel. Now, if an examination be made of the ordinary diet of persons who eat bread and potatoes, and more or less of other vegetables, it will be seen that the per cent of tissue-forming elements according to practical, instead of theoretical standards, is not much too low."

"Then, if it is not needed in the system, what becomes of it?"

"It must be converted into tissue, heat, or be excreted, and as many have a tendency to eat too much heat or force-producing food, the surplus cannot, in such cases, be converted into heat, but must be excreted from the system."

"In what way?"

"By the kidneys in the form of uric acid and urea."

"How about the savages who live almost exclusively on meat?"

"There is no doubt but what their strength and endurance was of a very high character, but that is easily explained. The Indian, in his original savage state, was not cursed with the frying pan, nor was he handicapped by hereditary weaknesses common to the frailties of civilization. But these are not the principal reasons. It was his out-door life, roaming over forests, mountains and valleys, that gave him a vigor of constitution which made it possible to live on any kind of a diet which furnished the necessary nutriment."

"Then injury from meat is not apparent, provided it be wholly used up in the system?"

"That is it; but the fact that it isn't used up makes us deal with conditions just as we find them."

"What is the consequence when the surplus of meat is not used up and has to be thrown out of the system by the kidneys?"

"Well, that may go on for a considerable time without any apparent injury, while in many people, some disorder would be noticed at once. It has been learned by one of the greatest physicians of England that headaches, asthma, rheumatism and many other of the common ailments, are, to a large extent, due to the defective excretion of nitrogenous waste matter from excessive use of meats."

"How is it that trainers for athletic contests use meat almost exclusively?"

"No person having regard for the truth could fail to speak of these matters as they are, and it is not my purpose to advocate any food merely to support a theory."

"Then you recognize that trainers have gotten good results from a meat diet?"

"There is no doubt of that, any more than that good results have been obtained without meat."

"But, Doctor, you say that the evils resulting from meat diet are great. How do you harmonize that with what you have said about the diet of athletes?"

"That is easy enough. The conditions under which a

prize fighter is trained are very different from the ordinary individual."

"In what way?"

"Well, for a prize fighter, the greatest care is taken in the selection and preparation of his foods, and the food is supplied in amounts exactly suitable for his condition. Then in addition to this, the great amount of physical exercise burns up or uses up every bit of food taken into the system, and besides the exercise, the baths and massage make the skin very active in eliminating the effete tissue. Such conditions cannot be compared in any way with ordinary living."

"You spoke of the athletes having their meat very carefully prepared."

"The greatest care is taken in cooking meat for a prize fighter, and it is usually done in this way: Three choice steaks are cut, placed together and put on the broiler; the first coming in contact with the fire until it is cooked, and then the three are turned over so that the top steak is brought in contact with the broiler. The middle steak is cooked from the heat of the other two, and besides absorbs more or less of their juices."

"Then the prize fighter eats the middle steak?"

"Yes, he gets the best and the rest is either thrown away or fed to admiring animals not in training."
cooking all meats?"

"Why not apply this principle, as far as possible, to

"It should be. There is no article of food so badly cooked as meat. Tender, raw meat is, comparatively speaking, easily digested; but meat cooked until it is solid, especially if it be fried, is very difficult to digest, and sometimes seems absolutely indigestible."

"Why is it worse to fry it than to cook it in other ways?"

"Well, a very important part of the digestion of meat

should take place in the stomach, while fats are not digested there at all. Now, to coat the food that the juices of the stomach act on, with fat, amounts to smuggling it through the stomach without digestion. Of course, this is not absolutely so, but the tendency is strongly in that direction, especially for persons of weak stomachs. This explains why lean meat that is fried until it is hard, is such a fruitful source of dyspeptic troubles."

"How should meats be cooked?"

"Well, if soup is desired, the meat should be covered with cold water and gradually warmed and then stewed at a low temperature until the meat is sufficiently tender. The way to make a roast tender, is to first immerse it in boiling water and then put in an oven and roast at a low temperature."

"What is the object in this?"

"The object is to coagulate or sear the entire outer portion of the meat so that no juices can escape."

"Can this be done in any other way?"

"Yes, by putting it into an oven that is very hot, or by enveloping the meat in a layer of dough, which accomplishes the same result. The whole object is to cook the surface of the meat and form such a coating as will not allow either heat or juices to escape. Then, after the surface is treated in this way, the meat should be cooked at a low temperature. This will make it much more tender."

"How about cooking steaks?"

"The same principle applies to steaks. The only way to cook a steak is to broil it—frying is abominable. Steak should be cut thick and put on a broiler—charcoal preferred—and cooked on one side and then turned over and broiled on the other quickly, so as to preserve the natural juice of the meat. It may be skillet-broiled the same way."

"Then you would add butter to it afterwards?"

"Well, the practice of adding butter (often strong at that) would seem to be a very bad one, because it is always objectionable to heat butter, and the flavor of the meat is quite equal or superior to the flavor of the butter. If butter be added while hot, it is about as objectionable as if fried."

"Beef tea is a preparation extensively used for the sick, but as ordinarily made it has very little nutritive value."

"What is wrong with the ordinary methods?"

"In order to get the solid matter out of meat, it must be macerated in cold water. If a great deal of heat be applied, it simply coagulates the proteid elements and makes them solid, and keeps all the valuable part sealed in the particles of meat, instead of dissolving in the water. The meat extracts of commerce are made by chopping the meat into fine particles, and then adding sufficient amount of cold water to soak thoroughly. Of course more of the solid matter would be dissolved, if the meat is occasionally bruised a little. After it has stood for some hours it is pressed so that as much solid matter as possible is gotten out of the meat in this way. Where no press is at hand, the macerated meat may be put in a coarse cotton or linen cloth and the juice squeezed out with a lemon squeezer. It should then be cooked at a low temperature and flavored to suit. Not much cooking is required.

"Beef broth is made by stewing beef bones and gristly substances with portions of meat. If they are first soaked in cold water, and cooked at a low temperature, the water will absorb much more from the bones and the soup be much richer and more palatable."

"How do you make meat powder, or scraped meat?"

"One way is to scrape the meat of a tender beef-steak that has been broiled according to directions heretofore given. The small particles that are gotten out of the steak with a dull knife or spoon, are quite nutritious, while no considerable amount of coarse matter is taken up in this way.. Another way to treat meat is to chop it fine, cook and dry by slow fire for several hours and then grind it in a mill until it is reduced to a fine powder. All of these methods of treating beef may be useful in typhoid fevers, or even in lingering illness of any kind."

"Are there not many parts of the animal used for food besides the flesh?"

"Yes; the pancreas, thyroid gland, what is known as the third stomach of the cow, called tripe, the heart, liver, kidneys, brains and sometimes the blood."

"What is the sweetbread?"

"Strictly speaking the sweetbread is the thyroid gland, although the pancreas is known and sold by that name."

"Have these any food value?"

"The sweetbread is said to have considerable value, is easily digested and is supposed to have a great deal of merit in regulating certain disorders of nutrition."

"What about tripe?"

"The Germans eat tripe, but not many Americans. It is very similar to meat, but is more easily digested.

"The heart is considered very tough and undesirable as a food, although it is very rich in tissue-forming material."

"Doctor, I suppose the liver is more extensively used as food than any other organs of the animals?"

"Yes; a great many people are fond of liver, and it is much more tender than either the heart or meat, but is less nutritious. If the liver and the kidneys are cooked for a great length of time, they become tough and difficult to digest. Many people are fond of animal brains. The constituents are very similar to that of eggs, only perhaps somewhat richer in fat. The blood is not used to any extent in this country, but it is used as a food in foreign countries. Tongue is largely fat and is about as hard to digest as fat pork."

"As gelatine is an article of commerce, I suppose the people would be glad to know something about its properties. What can you say of it, Doctor?"

"Gelatine is an important part of bones, tendons and ligaments, and it is from these that the gelatine of commerce is manufactured. It is somewhat different from meat, and will not of itself support life, but it is a very valuable food and is easily digested. The Keystone gelatine, made by the Michigan Carbon Co., is much superior to gelatine formerly sold."

"Is it not used extensively in making jellies?"

"Yes, it is. Ordinary jellies that are manufactured and sold through the trade, are mainly gelatine colored and flavored, and very often with essential oils instead of fruit flavors. They are much more easily digested than homemade fruit jellies, but much less palatable."

"How does veal compare with beef as a food?"

"One would naturally suppose that veal would be very much more easily digested, and in every way superior to beef, because calves for veal are young and tender."

"How does the composition compare with that of beef?"

"Veal contains considerable less fat than beef; otherwise, the per cent of tissue-forming substance is about the same as that of round steak, but is not so easily digested."

"Why is it not far more digestible, being so much more tender?"

"That is a matter in which authorities do not quite agree. Veal is much more favored in Europe than it is in this country, and the only reason that can be given why veal should disagree with people is because of the closeness of its texture, and it is probable that it is due to this fact that the digestive juices of the stomach do not penetrate it as quickly as ordinary beef or any other fresh meat."

"Doctor, you speak of mutton as though it was a common article of food. It may be that you have not boarded much, and therefore are not acquainted with the fact that it is lamb that is universally used and not mutton; at least it is always lamb on the bills-of-fare."

"One would think that lamb was too innocent a subject for use in perpetrating a fraud."

"In what way does mutton differ from beef?"

"It does not differ materially from beef except in flavor. It is supposed to contain more fat and less mineral substance, but the difference is not great. Mutton is about as difficult to digest as beef, although the fat is still firmer and more likely to disagree than the fat of other animals."

"At what age does the sheep make the best meat?"

"It is said that animals at least three years old make the best mutton, and that the main reason for the superiority of mutton produced in England is due to the greater age of the animals. Mutton is said to be somewhat constipating, but it is doubtful if it is more so than other lean meats."

"Venison from a young deer is believed to be the most palatable and easiest to digest of all meats; but as few people have an opportunity to eat venison, its composition is not a matter of great importance, although it is very similar to that of lean beef."

"Doctor, why does meat spoil so quickly?"

"That is partly due to the blood that is in it. The prevailing method of slaughtering animals is to shoot them or strike them a blow in the head. This is a good method for the butcher, but not for the meat."

"Why so?"

"Because the shock paralyzes the body and keeps in most of the blood, thereby increasing the weight of the meat. If animals are bled to death the meat is superior in quality and will keep longer and it is really a more humane way to slaughter them."

CHAPTER XV.

PORK.

"Doctor, you are not going to forget the hog, are you?"

"No, I see him too often for that; but I guess you mean pork."

"Yes, I mean the porker."

"Pork probably forms the most important part of a meat eater's diet; at least, more people eat pork than beef."

"A great many people say pork is not fit to eat?"

"Yes; some people say the same thing of beef, while there are others who say that neither is fit to eat."

"But why should there be more prejudice against pork than other meats?"

"Well, it may be that some people believe that the devils that were driven into the swine are still there, but most likely the prejudice to pork is because of the amount of fat it contains."

"Does not fat have as much place as lean meat for a proper diet?"

"The needs of the system for fat are certainly just as urgent as for lean meat, but from the fact that ordinarily it does not require as much fat as lean, it is for that reason supposed, by many, to be unnecessary. Then again, many people take the fats needed in the form of butter, cream or oils."

"What real value has fat for food?"

"Well, that will be treated under fats and oils. This much, however, might be said, that fat for many people is somewhat nauseating."

"Why so?"

"That is exceedingly difficult to determine, but one thing is well known, that when consumption has once taken hold, it frequently happens that one of the first and most noticeable signs is repugnance to fat. Just why it should cause nausea, is difficult to say."

"Don't fat agree with some people better than lean?"

"Yes, it does. Some people have good pancreatic digestion and can therefore eat fat meat and starches without the slightest feeling of discomfort, while the same person may not be able to tolerate any lean meat without feeling great distress, such as weight in the stomach, etc."

"What effect has the fat meat as compared with the lean?"

"Well, the primary or first use of lean meat is to supply tissue, while that of fat is to supply heat, although fat enters into many tissues, but does not form the frame-work or connective part."

"Then the reason that pork is more difficult to digest is because of the excess of fat it contains?"

"That is probably the principal reason, although pork is a firmer meat than beef, and is, therefore, naturally more difficult to disintegrate."

"How about the composition of pork?"

"Well, it. does not differ greatly from ordinary beef in amount of tissue-forming food, but has from double to three times the amount of fat, but less water."

"The question naturally arises whether salt meat or fresh meat is the most healthful?"

"That is easy to settle so far as digestion is concerned. Salt in considerable amount itself retards digestion. Besides that, meats that have been heavily salted become very firm. It follows then, that if salted meat be fried, the salt, together with the process of frying, makes the meat almost or entirely indigestible. The objection to fried

meat has already been explained, but perhaps it ought to
be repeated so many times that the people would get tired
of seeing it, because there is no one article of food which
does so much mischief as fried meat.   Fat bacon not included."

"How then would you treat bacon?"

"Bacon, if all fat, is comparatively easy to digest, and
would not be more difficult than any of the other fats,
and while frying is objectionable for bacon, it is far less
so than for lean meat.   Broiled bacon is comparatively
easily digested, and if fat food be needed, it is well-nigh
as valuable as butter or cream."

"Suppose ham or pork be boiled?"

"Boiled ham is as good as other meat, for the boiling
takes up a considerable portion of the salt and makes the
meat much more soluble; in fact, there is no comparison
between boiled and fried ham.   In certain diseases of the
stomach, boiled ham is the most useful of all meats."

"How about the composition of ham?"

"It has about the same general composition as pork—16
or 18 per cent. of tissue-forming food; and 35 to 40 per
cent of fat."

"Is fresh pork a good article of diet?"

"It might do very well for people who have been around
the world four or five times."

"But not many people have been around the world that
many times."

"That's the point."

"Doctor, I suppose that chickens and turkeys stand first
among common meats?"

"That is true; for the chicken is always an easy victim,
and can be caught and forced into the pot after company
comes."

"But that don't make it good. What is its value as food?"

"Of all the meat foods, it is the richest in tissue-formers. It contains only a very small per cent of fat, is not very tough, and there are no unusual difficulties or objections to it. We can therefore say that on the whole, it is the most digestible of all the common meats. The dark meat of a chicken is richer and more difficult to digest than the white."

"What is the difference between the chicken and the turkey?"

"The turkey contains more fat, but both contain on an average nearly twenty-five per cent of tissue-forming food. The chicken, ordinarily, has only three or four per cent of fat, while the turkey has eight or ten."

"Are chickens and turkeys different from ducks and geese?"

"Not very different, except that both ducks and geese contain more fat. This is especially true of the goose, for it has been known to be more than one-third fat. Fowls like turkeys and geese, containing a large amount of fat, are less digestible than chicken."

"How does wild game, such as pigeons, quails and partridges, compare with chicken?"

"The properties are about the same, only as a rule they are more tender."

"How should these various kinds of fowls be cooked?"

"By stewing or roasting. It is just as objectionable to fry chickens as it is beef, for the same reason; that is, that the fat, to some extent, prevents the action of digestive juices in the stomach."

"Doctor, fish is ordinarily considered a much lighter diet than meat; is there good reason for this belief?"

"Well, there is an occasional person with whom fish

does not agree, although fish is far less objectionable as an article of diet than meat."

"On what ground?"

"Because fish is much more easily digested—their fiber is shorter. They contain ordinarily only a small per cent of fat, a considerable amount of phosphorus, and do not produce the many ill effects resulting from uric acid tendencies common to an ordinary meat diet."

"How does fish compare with beef or pork?"

"Fresh fish has about the same amount of tissue-forming substance as good steak, but ordinarily not more than half as much fat."

"Then you strongly recommend fish as an article of diet in place of meat?"

"It would be far better for meat eaters if they ate less beef and pork, and more fish."

"Are there any diseases in which fish have particular value as food?"

"Yes; in diseases of the kidneys, such as Bright's disease; also in gout and other diseases."

"How should fish be cooked?"

"Broiled or baked. It is just as objectionable to fry fish as any other kind of meat."

"Are there any people who eat reptiles?"

"None but savage races, although the Europeans and Americans and other races are fond of turtles, which really belong to the general class of reptiles."

"Is the turtle a good article of food?"

"It is very similar in composition to chicken, only the oil and the flavoring matter is more pronounced, and for those who like turtles, this is probably the reason why they prefer the turtle to most other kinds of meat."

"Are there no other kinds of food that should be classed with meats?"

"Yes, there is probably no food in any class that is so universally liked as oysters. As the Irishman facetiously remarked, 'the oyster is the favorite American bird.'"

"Why is this?"

"Doubtless because of the richness of its flavor. There is nothing extraordinary about it otherwise."

"Some people believe oysters to be very nutritious."

"Pound for pound, they are only about half as nutritious as beefsteak, and not more digestible, unless eaten raw.

"Why better raw than cooked?"

"For this reason: When an oyster is stewed a portion of it becomes quite tough; instead of being easily digested, it is difficult."

"Are they not better when roasted?"

"If they can be roasted in the shell and only lightly cooked, they should be almost as digestible as when eaten raw."

"To what class of foods do they belong?"

"Tissue-formers. They furnish very little fat or fuel for the body and should only be eaten with bread, potatoes, or cereals of some kind."

"How should they be cooked?"

"Roasting is preferable. If not convenient to roast, they may either be baked or stewed, but never fried."

"What about eating them with various kinds of pickles?"

"The pickles would be very likely to disagree if the oysters did not. If anything sour is desired, they should be eaten with lemon juice."

"I have heard of cases of poisoning by oysters."

"Yes, that sometimes happens when the water around the oyster bed has been contaminated. They have been known to cause an epidemic of typhoid fever, but they are not so likely to cause disorders of digestion as the lobster, or crab. Shell fish are scavengers, and many urge

that their use be discontinued because of the many cases
of poisoning produced by their use."

"The lobster is the most likely to produce illness?"

"Yes, lobsters are more likely to disagree with weak
stomachs and cause violent attacks of indigestion than almost any other food. All shell-fish, and other fish too,
for that matter, seem to be particularly bad when tainted
with decay, so that there is hardly any substance which we
could eat, more likely to poison than tainted fish, whether
it be canned salmon, oysters, lobsters, or any kind of fish
whatever. Everything in the way of fish is better if used
fresh. They should be carefully kept in ice in warm
weather."

"Doctor, I suppose you will attack the egg with great
vigor, and probably shell it out of its house and home?"

"You have hit it eggsactly," said the doctor, "although
at times it takes a great deal of courage to face it."

"Well, Doctor, I like to face them if fried with good
ham."

"Ham and eggs is a favorite combination, but a prolific source of dyspepsia."

"I thought eggs were the easiest of all foods to digest."

"Yes, that is the general belief. A raw egg is, but a
fried one belongs to the class of never or forever."

"Why are fried eggs so bad?"

"Because every bit of heat that is applied to an egg
makes it harder, and when it is fried for a time, it is
very much like leather. Then if to this condition fat be
added until it is thoroughly saturated, it becomes as difficult to digest as hard fried ham, and the two together are
enough to send anybody to the doctor."

"I have always understood that the egg was very nutritious?"

"So it is, but it has been overestimated by a great many

people. It furnishes all the necessary elements for the life of a chick, and has therefore all that is necessary to support the life of an individual."

"Then eggs ought to be one of the most useful of all the articles of diet?"

"Yes, they are useful; at the same time, there are some objections to them."

"What are they?"

"They contain almost the smallest amount of waste matter of any food, and are therefore constipating. A great many people dare not eat them on this account."

"Is there no way of overcoming this difficulty?"

"There is no way of changing the egg. The only thing that can be done is to eat them in small quantities, say one at a time with food containing a large amount of waste matter, such as the cereals with part of their bran, or with coarse vegetables."

"Doctor, I have often known people who were told not to eat meat, and they thought they were not disobeying when they ate eggs."

"In that case, they kept the letter of the command, but not the spirit, for practically eggs are the same as meat."

"In what way do they differ from meat?"

"Well, eggs have about fifteen per cent tissue-forming substance, and twelve per cent fat. This is only a trifle below that of ordinary steak, with which they favorably compare.

"Which contains the least waste matter, meat or eggs?"

"Meat is less constipating than eggs, although if there be a tendency to headaches and what is known as uric acid condition of the blood, eggs are much less objectionable than meat."

"Then eggs properly go with a vegetable diet?"

"Yes, meat and eggs make the diet too strong on the side of tissue-forming food."

"Have eggs any other special value?"

"They are good for a quick lunch, or rather a drink. One or two raw eggs with the juice of half a lemon makes an admirable drink, and if one is greatly crowded for time there is nothing more suitable than egg lemonade."

"Why is this?"

"Because it does not require any mastication, and the acid helps digest it. They are easily digested, and no injury results, because they are swallowed in a hurry, which would not be the case with any other food."

"Since you say that eggs must not be fried, I suppose you advocate that they be boiled or poached?"

"Eggs may be boiled, poached, roasted or baked (called shirred), and you can lay down a general rule that the less an egg is cooked, especially the white, the better, although the yolk will stand cooking until it becomes mealy."

"With the exception of milk, eggs are the best for feeding the sick, and are sometimes better than milk. Albumen water is made by stirring the whites of eggs in water. A pinch of salt and a little flavoring may be added. Equal parts milk and egg is much more nourishing than milk alone. The yolk of eggs is richer than the white, and should be used largely where a very rich diet is necessary, as in consumption, anaemia, and other diseases."

"Doctor, a great many people do not distinguish between fat and flesh."

"That is true, but there is a great difference."

"Will you kindly explain it?"

"Fat is both fuel and covering for the body, and before a person dies of starvation about eighty or ninety per cent of it will be used up. The principal use for fats when taken into the system is for these purposes."

"There is not much waste matter then in them?"

"No, there is practically none at all. Fats that are taken into the system as food, if absorbed, are either stored, or burned up in the production of heat and force."

"How much of the system is composed of fat?"

"An average person is supposed to be about one-fifth fat, although many people have a much higher per cent."

"You say that fat is used as a covering for the body?"

"Yes, fat prevents the radiation of heat, and this is the reason why a fat person usually eats less and requires less clothing to keep warm."

"What kind of foods produce fat?"

"All the fats and oils, both animal and vegetable, together with starch and sugar."

"Has fat any other use?"

"It is of use in giving persons a comely appearance, and also for storing energy, so that in the event of illness or deprivation of food, life can be sustained for a considerable length of time without any food at all. I suppose that most persons are familiar with the fasting experiment of Tanner, who lived forty days without taking any food except water."

"Is there any material difference between the fats and the oils?"

"Only a slight difference. They have essentially the same composition whether they be animal or vegetable oils."

"Why is it that fats disagree with so many people?"

"That is quite difficult to answer, but it is believed that when a considerable quantity of fat is taken with the food, that it coats the food and prevents the action of the digestive juices in the stomach, very much as it does with fried meats, although most persons can eat a fair amount of fat taken as butter or cream. Such fats as butter, lard, and oil, are called free fats. The fat globules are not held together by any tissue. Free fats are much more likely to cause indigestion than emulsified fats, or when in the form of fat meat or powdered nuts."

"What is the objection to fried fats?"

"Heat bursts the fat globules, and the fat being to a certain extent burned, a chemical change takes place which makes an irritating fatty acid."

"Then that would affect lard, would it not?

"Well, lard can be made at a very low temperature if it is done properly, and frying it out or rendering, so called, does not necessarily make it more indigestible than other free fats."

"Is not lard used much too extensively in cooking?"

"Well, it is very much better for ordinary use than butter, but at the same time frying almost any kind of food is not in harmony with good living, but is very much worse for some kinds of food than others. Dyspeptics should use free fats very sparingly, if at all."

"What kind of fats are tallow and suet?"

"Tallow is ordinary beef fat, and suet is the kidney fat of beef."

"Are there not many mixtures of these used under various names?"

"Yes, they are mixed with cotton-seed oil—possibly other kinds of oil,—and sold extensively for the same use as lard."

"How is oleomargarine made?"

"Oleomargarine is beef fat treated with a few chemicals mixed with a small amount of butter and sold for butter."

"I did not know that it contained any butter."

"The ordinary formula for making oleomargarine does not include butter, but it is sometimes put through a process that they call churning with milk to give it a flavor of butter. The different compounds of oleomargarine and butterine are made in different ways, but are substantially the same product."

"Are they healthful?"

"They are better than poor butter, but being somewhat more solid, are a little more difficult to digest."

"Doctor, you said a moment ago that cotton-seed oil is used?"

"Yes, the manufacture of cotton-seed oil from cotton seed has grown to be an important business. It is refined and as already mentioned, is then mixed with other fats for cooking purposes."

"How is olive oil made?"

"Olive oil is made from very ripe olives. It is used principally as a table oil for salad dressing. It is claimed that many other kinds of oil are sold under the name of olive oil."

"What other kinds of oil are used for food?"

"Cocoanut oil, peanut oil, and cocoa butter, the latter being made from cacao seeds used in the manufacture of cocoa. These various oils vary in flavor, and slightly in composition, but are used for the same purposes."

"Doctor, you forgot cod-liver oil, did you not?"

"While cod-liver oil is a food, it is usually prescribed as a medicine."

"Why is that?"

"The cod-liver oil contains some chemical elements not found in other oils, but it is probable that it is often prescribed because heretofore no other oil suitable for administration was readily obtainable."

"For what purpose is cod-liver oil prescribed?"

"To get an oil that is easily absorbed in the system. Persons afflicted with consumption or wasting diseases

have a continual tendency to grow thinner. This is because the system does not take up and absorb enough
heat-producing material to prevent the destruction of the
tissues of the body for heat production; or to make it
plainer, the system must have heat, and when not furnished by the food, it burns up its own tissues until the
system wastes away, and it is to prevent this wasting that
consumptives take cod-liver oil and other fats,—cod-liver
oil being preferable, because more readily absorbed."

"What kind of a product is glycerine?"

"Glycerine is sometimes described as the sugar of fat.
A very poor description, but gives a faint idea of its
character. It is the part of fat which does not readily
saponify in the manufacture of soap.   It is not used to any
considerable extent as a food."

"Are not fats and oils frequently given as cathartics?"

"Some oils are used for that purpose.   The tendency of
all fats and oils is to be slightly laxative."

CHAPTER XVII.

FRUITS

"Doctor, I cannot remember a time when I did not hear about the healthfulness of fruit, and yet it is claimed that it often makes people ill."

"Yes; the people believe that fruit eating is conducive to health, and the whole race, it seems, has been disposed to follow Adam who ate the apple against the commandment."

"If fruit be so healthful why don't those who are ailing (and that includes a large per cent of the people) eat it and get well?"

"Your question assumes a great deal, and is therefore hard to answer. Probably the difficulty is in being able to determine the proper use of fruit."

"It seems strange that the learned doctors of the world should not have found this out in several thousand years."

"That is not so strange after all; for it should be borne in mind that very few persons in any profession make any extensive original investigations, and it is only due to the modern achievements in chemistry, which enable us to analyze the secretions and excretions of the body with the digestive processes, that has thrown any light on the uses of fruit."

"I was under the impression that a good many theories had been advanced why fruits were particularly beneficial in promoting health and curing disease."

"So there have. It has been urged by writers on diet that fruits assisted in burning up the starches, and in the production of heat; by others that it was the mineral salts

175

that made them valuable. These are not all the theories, for they have been numerous."

"Is there no truth in these theories?"

"As to the first it is the opposite of truth, because acids and starches are entirely incompatible, so that fruits containing acids cannot help the digestion of starch, sugars or fats; neither does it directly help their absorption or the process of burning them up."

"It is generally believed that fruits are laxative. There must be some truth in this?"

"Well, not all of them at least."

"Then according to your views no correct explanation as to the use of fruits has ever been made."

"No; the proper uses of fruits are still unknown to the laity, and only partially known to the medical profession, but as no satisfactory explanation has been given for their use, it is not clear when they are useful and when likely to be injurious. Their palatability has probably caused their popularity to a greater extent than their efficacy in a medicinal way."

"Then that accounts for the fact that their use sometimes does harm, while at other times seems to be just what the system required."

"It is this way; whenever we guess at a thing we are strongly indorsing the apple?"
more liable to be wrong than right, although we may happen to know something about it."

"Then, Doctor, if you have found out all the uses of fruits, you can do a great deal toward enlightening the world."

"I can hardly make a claim of that kind, and I didn't mean that I had learned all there is to be known on the subject. My remark was more in the nature of a lament, because so little was really known."

"At any rate, Doctor, you have my curiosity aroused to a high pitch to know what you regard as the principal uses of fruit?"

"Well, in discussing the question of digestion, I told you that the entire digestive tract from the mouth downward was lined with a soft membrane, called mucous membrane. Now, the one great, fundamental use of fruits is to cleanse the mucous membrane."

"Then all the other uses are secondary to this?"

"I would hardly put it in that way, but rather say that the benefits usually ascribed to fruits were incidental to the cleansing of this mucous membrane."

"That is not very clear."

"Then I will explain."

"One of the effects said to be due to fruit eating is that it is laxative."

"Then how does that result from the cleansing of the membrane?"

"When the membranes are cleansed the secretions are better able to perform their functions, and besides that, it causes food and detritus and mucus to be removed."

"How does it do that?"

'The fruit acts on the membranes, probably killing the bacteria, and when that happens, they naturally pass away. Then it acts in another way. It increases the specific gravity of the urine, i. e., the weight, which makes it possible for it to carry away more of the solids of the body waste."

"Then the removal of effete tissue, mucus and excess of nitrogenous foods not used in the system, is helped by increasing the weight of the urine?"

"Yes; by increasing the eliminating capacity of the kidneys."

"Can you demonstrate what you have said about the effect of acids?"

"Yes; you can do that in part for yourself."

"I should like to try."

"Take a lemon and suck the juice and allow it to come in contact with as much of the membranes of the mouth and throat as possible. If there is any accumulation of mucus, that the acid touches, it will be removed."

"I have often done that, but I never thought of that as being the chief action of fruits, but I can readily believe from my own experience that what you say is true."

"Yes it is true, and I have demonstrated the other fact so often that I can positively say that fruits increase the specific gravity of the urine."

"To what is its action due?"

"Partly due, no doubt, to its power to destroy bacteria and partly to the fact that the acids by contact with mucous membranes stimulate them to activity. Anyone with a coated tongue who will eat something like sour apples or lemons will have the coating quickly removed.

"Then under what conditions are fruits beneficial?"

"Whenever there is torpidity of liver or congestion with mucous secretions and whenever there is excessive alkaline fermentation in the bowels." (See diseases of liver and intestines.)

"When are fruits harmful?"

"Whenever there is an excessive acid condition of the stomach, intestines and urine."

"How are we to determine these different conditions?"

"Excessive acidity of the stomach is indicated by one of the most common expressions, 'heart-burn.' This is really not a condition of the heart as one would suppose from the sensation this feeling gives, although it seems directly in the region of the heart."

"Yes, I have often experienced such a sensation, at times amounting to absolute pain, and have been alarmed, thinking it might be a disease of the heart. Now, Doctor, can you not give me a clear explanation of what this is, and in what manner it is brought about?"

"I will try to do so, as it is one of those conditions which while not necessarily harmful, causes much uneasiness and alarm at the same time, and if the 'heart-burn' be continuous, it is the forerunner of serious stomach trouble."

"Then 'heart-burn' is the result of irritation of some part of the stomach?"

"Yes, the burning sensation is at the upper part, called the esophageal end, which is parallel and within an inch of the heart; hence the error in supposing that it came from the heart. You will better understand how the burning sensation occurs, when I explain that the esophageal end of the stomach lies in folds something like the gathered end of a tobacco pouch. Now, when there is an excess of acid in the stomach with gaseous distention, the folds at the esophageal end are stretched out, which leaves the entire surface exposed to the irritating influence of the corroding substance in the stomach. It is this that causes the burning sensation called 'heart-burn.'"

"May there not be an excessive acidity without 'heart-burn?'"

"Certainly. Acid eructations or uneasiness two hours after eating food, show too much acid; also excessive secretion of gastric juice (as distinguished from excessive acidity from fermentation) comes within the principles excluding the use of sour fruits."

"Will you explain the symptoms of excessive gastric secretion?"

"That will come under diseases of the stomach (see excessive or hyper-secretion, page ....)."

"Doctor, is it not a fact that heart-burn often results because of torpid liver and inacitivity of the bowels?"

"Yes."

"Then acids are indicated for the liver, and contra-indicated for the stomach. How do you harmonize the two?"

"In such cases it may be well to neutralize the contents of the stomach by an alkali and then not eat the usual evening meal. By kneading the abdomen for ten minutes before retiring, the stomach should be entirely empty by morning. Now if the juice of a half lemon be added to a teacup of moderately hot water, and be drunk without sugar an hour before breakfast, and the abdomen again kneaded to clear the stomach, it will be in good condition for a breakfast of a soft boiled egg and a little milk. This treatment is only for ordinary attacks of indigestion, resulting from torpidity of liver. Where the inflammation has been continuous and of long standing, fruit acids are harmful, because the membranes of the stomach are too sensitive, and acids increase the irritation."

"Is there any serious injury done by eating fruits?"

"Yes, great injury. Besides the acid condition of the stomach, where sour fruits are harmful, there are many acute attacks of diarrhoea, cholera morbus and similar complaints which are brought on by eating tainted, tough or green fruit."

"What do you mean by tainted fruit?"

"Fruit that has commenced to decay. As a rule, people have a keen relish for almost all kinds of fruit, and as the flavor is better in its natural state than when cooked, it is often served that way when it is really dangerous to health and life."

"Why is this?"

"Because a large per cent of fruits are partially spoiled before they reach the consumer, and if eaten without being cooked, violent disturbances of the digestive organs are likely to result. If fruit of any kind be spoiled in the least, it must be cooked sufficiently to arrest all fermentation; otherwise, it is unfit for use."

"What kind of fruits are the most likely to be spoiled?"

"Probably strawberries and peaches, but all fruits, such as berries of every kind, plums, pears, bananas, and even apples and grapes are sometimes tainted with decay."

"I suppose that a speck or so in an apple would not injure it any, if all the decayed part was removed?"

"An apple with a rotten speck is not fit to eat unless well cooked; no matter how sound a part of it may seem, the apple is contaminated."

"It would seem that people would learn better than to eat spoiled fruit?"

"But they do it, and it furnishes the doctors half their business or more, at certain seasons of the year, when fruit is moderately plentiful, and the weather favors decay."

"Does cooking stop all decay?"

"Yes, for a time, but it must not be understood that cooking restores rotten fruit, but when the decayed parts are removed, the cooking makes what is apparently sound, but merely contaminated, eatable."

"Then fruit is after all dangerous?"

"It ought not to be, but until the people learn that spoiled fruit begets familiarity with the doctor and the undertaker, the injury will go right on."

"When should fruits be eaten?"

"You ought to have added the purpose for which they are to be used; also the kind of fruit and the condition of the individual, for all these are modifying circumstances. To get the best effect, it should be eaten on an empty stomach, that means three or four hours after meals, or better still, an hour before breakfast. This of course applies only to acid fruits without solid matter. Rich fruits like figs and bananas should be eaten with regular meals."

"Does eating acid fruit on an empty stomach aid digestion?"

"Very much, and relieves constipation, provided of course, the condition of the patient is such as to indicate the need of fruits."

"Under what other circumstances would fruits aid digestion?"

"Well, fruits to a certain extent supplement the natural gastric secretion, especially in the digestion of tissue-forming food, such as meat, eggs, oysters, peas, beans and wheat gluten. These foods will be more readily digested when acid fruits are eaten with them, provided of course, the stomach is not already too acid."

"When are fruits indigestible?"

"When they are solid or tough. Green fruit is always more or less solid, and if it is pulled green, it may apparently ripen, but still be exceedingly tough. Fruit of this kind ought to be let alone, unless cooked until soft."

"This would strike fruits that are shipped long distances?"

"Yes, peaches and bananas and some other fruits that are shipped a long distance, and ripened in cellars or in boxes, are not suitable to be eaten raw."

"Boys seem to have a strange weakness for green apples, but the apples can hardly be said to have a weakness for boys, because they often lay them out.   Why is this?"

"Because they are tough and acrid.   They are not easily disintegrated, and therefore irritate the lining membrane of the digestive organs, and cause diarrhoea."

"Then when any kind of fruit is hard or tough it should not be eaten?"

"Not in its raw state, and not when cooked unless it cooks soft."

"Is it proper to eat fruit at meal times or between meals?"

"Fruits that are not very sour may be eaten at meal times with any kind of food.   Sour fruit when permissible at all may be eaten with meat, beans, peas, eggs, oysters, but not with milk, bread or vegetable foods containing much starch, such as rice, potatoes and oat-meal."

"Does not the use of fruit have a tendency to increase the consumption of sugar, which you say is too great already?"

"It does, but it ought not to be so, for there is no reason or need for covering fruit with sugar as many people do."

"But it is disagreeable to taste and disturbs some people's stomachs."

"As to taste, that is much a matter of habit, for most fresh berries and fruits are better as pulled, than with sugar, and as a rule it is the sugar that disagrees, or if not, it is the tough skin or seeds.   Of course, if acids are taken where there is already too much acid in the stomach, that would of itself be a cause for increasing the unfavorable symptoms."

"Is there any way of treating sour fruits so as to make them palatable without sugar?"

"Yes, if it is not desirable to use sugar, bi-carbonate of soda (common baking soda) may be added to sour fruits when cooking.   This neutralizes the excessive amount of acid. However, many of the sourest fruits like cranberries should not be used at all by some people, while others may use such fruits with sugar, without any apparent ill results."

"You mention skins and pits as causing injury?"

"Yes, all skins, seeds, especially seeds of any considerable size, should be separated, and never swallowed; but this will receive further mention when we discuss each particular fruit."

"Which of the fruits do you consider most valuable?"

"In the temperate zones, the apple. The apple is king of fruits, and one is tempted to say of it:

Blessed be thy crimson cheek,
  Kissed alike by the sun and the breeze;
So good, so beautiful, so divinely meek;
  There is none thy equal, on earth or seas."

"That is beautiful, Doctor, and shows poetic genius, but why does the apple so inspire you? I hope that you haven't been drinking apple jack?"

"Well, if the apple be worthy of a place in the songs of Solomon, who says, 'As the apple tree among the trees of the wood, so is my beloved among the sons; and then again he says, 'Stay me with flagons, comfort me with apples.' Surely if Solomon could say so much, why should not the doctor, who appreciates their excellencies, be equally enthusiastic in their praise?'

"Then you think Solomon showed his wisdom in so

"Yes, no doubt he could have spoken in still more eloquent terms, could he have tasted some of our Nineteenth century fruit."

"I suppose that one of the good traits belonging to the apple is that we have it all the year?"

"Yes, that is one, and a very important one. Another is that the apple is the least harmful or misused of all fruits; for it is seldom used with any serious injury; in other words, it is the least likely to be misused."

"What are the many other good traits that you ascribe to the apple?"

"Well, another excellent thing about the apple is the variety. There are some 300 varieties in cultivation, each different from the other in flavor, and varying from the sweetness of sugar to the sourness of the lemon, or nearly so."

"Doctor, what is the average composition of apples?"

"They vary greatly, ranging from about eighty-two to ninety per cent water. The food elements are principally gum and sugar—the sugar varying according to variety, but it usually runs from five to seven per cent. The apple ordinarily has very little tissue-forming element; it is strictly a heat producer, so far as you could consider it as a food."

"What about the waste matter of the apple?"

"It is only about two per cent including the skin and the core, and without these it would be very small, so that the apple is not laxative because of this, for nearly all the vegetables, and for that matter, nearly all foods, have a larger amount of waste matter than the apple has, exclusive of core and skin."

"There must be something else that makes the apple so valuable?"

"Yes, its particular effect results from the acids and mineral matter. There is usually about one per cent of malic acid, although of course among the numerous varieties this would vary greatly. The apple also contains considerable potash and soda, and a trace of lime, magnesium and iron. Some have reported that laborers could live on the apple alone, but we doubt this very much. In this respect, it has rather less tissue-forming

food than the potato.   Bulk for bulk, the apple is slightly
less nutritious than the potato, but its sugar and gum
compare favorably with the starch of the potato as a
heat producer; but of course the apple, on account of its
acids, has many uses entirely unknown to the potato."

"Have not a great many writers unduly extolled the
healthfulness of the apple?"

"Perhaps they have; it certainly will not correct all the
abuses that may be inflicted upon the system.   It is only
an aid when there is proper consideration as to the kind
and quality of food consumed, and the habits are otherwise good."

"What are the various uses of the apple as an article of
diet?"

"Probably the first is, that the apple adds variety to the
diet, for it can be cooked in so many ways, it can hardly
fail to revive a failing appetite.   Apples that are not very
sour may be used with the cereals to give them flavor.
By a little care in this way, many persons can be induced
to eat cereals, who would not otherwise eat them at all,
because they do not like them.   Its other uses depend
upon its acids."

"Then apple cider must be good?"

"Cider, being the juice of the apple, when fresh has
about the same uses as the apple.   Sweet cider has been
known to benefit cases of aggravated constipation, when
apparently nothing else would."

"Why is this?"

"Very likely the great quantity of water together with
the acid exerted a stimulating effect on the bowels without
undue irritation."

"Is not vinegar made of cider?"

"Yes, but the less said about vinegar or the less used
the better, although cider vinegar is very palatable, due

to small particles of the apple which gives it its flavor."

"What is the best way to prepare apples for food?"

"Apples are best either baked or stewed, though they may be cooked in other ways for variety."

"Are dried apples equally wholesome?"

"Dried apples are better than none at all, but not so good as the fresh, owing to the fact that water is evaporated. Bulk for bulk, dried apples are twice as rich as before drying."

"Doctor, awhile ago you spoke of the peach as being a source of digestive disturbances; would you condemn it because of this?"

"No, the peach is a very choice fruit, and probably more people like a good peach better than any other fruit."

"Why are they so often the source of disease?"

"Because they ripen in hot weather and will not keep. The peach should be eaten within a few hours after it is pulled, and if it is not, it may become tainted, and cause violent gastric disturbance or diarrhoea."

"Is there no way of preventing this?"

"No way except to get the people to understand that they must not eat stale fruit, but if they do it must be cooked, so that all bacteria may be destroyed and decay arrested."

"Why not can or dry peaches?"

"That is a good way. They can be kept very well all the year. Good canned peaches are almost as good as fresh ones. What is known as pie peaches, containing green and solid lumps and more or less of the tough dirty skin of the peach, should not be used by anybody."

"In what does the peach differ from the apple?"

"It does not contain as much sugar, but as a rule more gum. The principal part of the peach, exclusive of water gum.   The principal part of the peach, exclusive of water, is known as pectose, which is a sort of gum."

"What kind of acid is found in the peach?"

"Principally malic acid, the same as in the apple.   Good peaches are almost as nutritious as apples, but much more care is needed in using them, because of the liability to either be green and tough, or over-ripe and tainted with decay.   A choice peach not too green or too ripe, is one of the most delicious things with which nature has provided man."

"Doctor, a good many people prefer the pear to any other fruit."

"That is doubtless because of its sweet taste; otherwise, it is not so rich as many of the other fruits."

"How does it compare with the peach?"

"Well, in a general way, it has about the same amount of water (83%), but the pear has about twice as much sugar (8%), and half as much gum or pectose.   The food value, although different in character, amounts to about the same as the peach.   The pear has but little acid, and it may therefore be used with any kind of food, because there is not sufficient quantity of acid to even coagulate milk to any noticeable extent."

"What would you consider the best way to use the pear?"

"Aside from its use as a pleasant fruit, it makes the best preserves and jam of any of the fruits, or at least, it is highly prized for that purpose."

"Some people declare the pear to be laxative, while others say it is astringent."

"Some varieties are astringent, but the excessive amount of sugar in the pear sometimes causes an abnormal fermentation, and results in diarrhoea; then again, the pear is frequently tough and may cause disturbances on this account. The same care is needed in using pears on account of being either hard or tainted with decay, as that of peaches. Owing to their large amount of sugar they should not be eaten by any one subject to sour stomach."

"The quince is the most solid of all the fruits, and unless well cooked is not eatable at all. It contains a large amount of malic acid and a great amount of gum. When thoroughly cooked, many people prize it highly for its flavor. It is slightly astringent."

"Is there any other use for it than as a stewed fruit?"

"It makes a jelly of the finest quality."

"Doctor, I suppose the grape is almost next to the apple?"

"Probably considering its universal use, it certainly ranks high, and if not next to the apple, it ought to be considered at least one of the most valuable of all our fruits."

"What nutriment is there in the grape?"

"That depends much on the variety. Some grapes have much less water than others. A fair average probably would be about 80% water, the principal other ingredient, besides waste, being sugar. In addition to the sugar, the grape has considerable tartaric acid, and when we consider the seed and skin, it has a very large amount of waste matter, but with these out as they should be, the waste matter is small. The grape has not enough nitrogen in it to make this element worth mentioning, and like the fruits just discussed, it is strictly a heat-producing food. There is also considerable mineral matter, soda, potash, magnesia and iron, in addition to tartaric acid."

"I have heard very well-informed people say that grape juice contained very nearly the same elements as blood?"

"When they said that, no matter who they were, they

were talking rank nonsense, because the grape lacks a
great deal of furnishing the necessary constitutents of the
blood."

"What uses has the grape?"

"It is a wholesome and pleasant fruit, if properly eaten."

"How is that?"

"The pulp should be dissolved and no one should swallow either the seeds or the skin."

"Has the grape any particular value in disease?"

"Yes, it has great value, but this will be considered
under the head of disease."

"What about wine?"

"Wine properly belongs with spirituous liquors."

"Of what are raisins made?"

"Raisins are dried grapes, also what are known as English currants, are really only inferior raisins."

"Are raisins healthful?"

"They contain about the same properties as grapes, but
owing to their toughness and their seeds, they should be
cooked and thoroughly masticated, and any one who
gives raisins to small children, does so at the risk of
causing their death."

CHAPTER XVIII.

FRUITS CONTINUED.

PLUM.

"The plum is a nice fruit, makes most delicious preserves and jam."

"But Doctor, a good many people think plums very unwholesome?"

"Doubtless a good many disorders have been produced by the plum, because they are so often tough, acrid, and therefore unsuitable for food; but choice varieties of plums that have been ripened on the tree are both delicious and wholesome, provided of course, the tough skin is not swallowed."

"How do they compare with other fruits?"

"They are very similar to the peach, only as a rule they are more acid. They usually contain a little less sugar, and about the same amount of gum. The per cent. of acid in the plum is ordinarily about 1½ and nearly double that of the peach. Some varieties of the plum are quite astringent."

"Are not prunes some variety of plums that have been dried?"

"Yes, the prune is really a plum, but a sweeter variety than the ordinary Damson or Green Gage plum."

"What value has the prune as a food?"

"The prune contains a large amount of sugar, and it is supposed to be very laxative, but it has been much overrated in this respect. It has no properties to cause it to be more laxative than most other fruits, and careful observation will show that it is not so in practice. Prunes should be very well stewed, as otherwise they are unfit to eat.   The removal of the tough skin by straining the pulp adds greatly to their food value."

"Doctor, how does the cherry rank as a fruit?"

"The cherry is a favorite of many people, but it ought not to rank very highly, because a large per cent of it is tough skin and water, and it is rather strong in acid."

"Then you do not recommend the cherry very strongly?"

"No, I do not.   If the juice is fresh and used for making jelly, perhaps one could justly extol it, but it has a very thick skin, and a small amount of pulp, which leaves very little of the fruit suitable for use.   Like other fruits, the sugar and acid vary much according to the variety, although it is very similar to that of the plum."

"The apricot and the nectarine are very similar to the peach, but are not as rich.   It does not need any extended description, because it is so nearly like the peach."

"Doctor, it rather seems as though you had slighted our berries?"

"Well, the berries are in such great favor, especially the strawberry, that some enthusiasts have said that the 'Lord could have made a better berry than the strawberry, but he didn't.'"

"What properties has the strawberry?"

"It does not differ as much as one would suppose from other fruits.   It contains some more acid than the average apple, not quite so much sugar, and a good deal of waste material or cellulose.   The nitrogenous or tissue-forming element of the strawberry is proportionately higher than most of the other fruits.   Ordinarily, it is about eighty-eight per cent water."

"Is there any injury likely to result from using strawberries?"

"Yes, many persons are injured by using stale strawberries. They do not keep but a short time, and like other fruits, when tainted they should be cooked, but the tendency is to merely add sufficient sugar to hide their decay."

"Has the strawberry any action different from that of other fruits?"

"Yes, it is more laxative, because of the stimulating effect the small seeds have on the intestines, and if strawberries are used judiciously, they have very great value, as they come early in the season, at a time when their flavor and their acid is needed to clear the system for hot weather. Strawberries make very delicious jelly and jam. They should not be used with milk, because their acid coagulates the milk, causing it to form little hard lumps or clots. Strawberries are charged with being the cause of hives and skin eruptions, but only in people who have some peculiarity—probably an excess of uric acid in the system."

"Doctor, you spoke of the acids coagulating milk; what kind of acid does the strawberry contain?"

"The strawberry contains both malic and citric acid, also potash, lime and soda salts. It is therefore slightly diuretic as well as laxative."

"Is the strawberry used in any other way, except as it is picked?"

"Well, it may be cooked, and used for flavoring other foods."

"The raspberry is one of the most palatable of the summer berries, but it is so much like the blackberry and blueberry they may all be discussed together."

"In what way are these different from other fruits?"

"They differ in this; they contain more seeds, or at least larger ones, and less water, and instead of being laxative as are most fruits, they are astringent, and wine made of blackberries is one of the most common remedies for diarrhoea or summer complaint."

"What properties have these berries as food?"

"Aside from their acid, and mineral salts, which are similar to those of the strawberry, they contain little, except sugar and their agreeable flavoring matter, common to various other fruits."

"To what do you ascribe their astringency?"

"Tannic acid, or something equivalent to it."

"What other berries besides the raspberry, blackberry and dewberry are astringent?"

"The elderberry and blueberry. The elderberry is not extensively used, although it makes an agreeable wine, and is made by many people for home use. The blueberry is a berry of commerce, of which there are several varieties. It has an agreeable flavor, and is not very pronounced in its action, because it contains little but seeds, sugar and flavoring matter."

"Cranberry, gooseberry, and currant, are all popular fruits. The cranberry is more extensively sold than any of the others."

"Why is this?"

"Because it matures late in the season, and is easily kept all winter."

"What are its properties?"

"Malic and citric acid in large amounts, a little flavoring, and an exceedingly tough skin."

"Can you recommend its use?"

"On account of the strong acid and skin of the cranberry, it disagrees with most people. If used at all, it should be stewed and strained so that the tough skins come off. This would practically make a jelly of it. The acid is exceedingly acrid, somewhat astringent and of rather doubtful use. It is sometimes useful as a disinfectant for inflammations, and is usually applied as a poultice."

"The gooseberry is much more favorably known in England than this country, as it requires a cool, moist climate for good fruit. It also has a tough skin and large seeds, and nothing to recommend it except its flavor and the sugar its contains. When green, it is very sour, but when fully matured and ripe, it contains quite a large per cent of sugar, more in fact than most other fruits. The currant is another tough-skinned fruit with large seeds. It does not differ greatly from the gooseberry, except that it never has so high a per cent of sugar. None of these berries should be used with their skins, and they are therefore more suitable for making jams and jellies, than for any other purpose."

"The mulberry has never been so extensively grown as its flavor would certainly warrant. There are few berries as rich as the mulberry, and it ought to have been planted everywhere, instead of the cherry, although it does not produce so large a crop nor is it so sure to bear."

"What are its properties?"

"It is very rich in its flavoring matter, has a high per cent of sugar, and contains about one and a half per cent of tartaric acid, and is therefore more like the grape than any other berry. It also contains considerable potash."

"Doctor, what is the leading fruit from the tropics?"

"Well, it is difficult to say whether the banana or lemon."

"For a food, which is best?"

"The banana. It is the only green fruit extensively used in this country, upon which life can be sustained for any length of time."

"Then, the composition will be interesting?"

"Yes, the banana contains one per cent or two per cent

of tissue-forming food, or about one-eighth that of entire wheat flour. It contains quite a large amount of gum and sugar, amounting in all to about fifteen per cent. The banana contains less water than most other fruits, being only about eighty per cent water, while most of the others range from eighty-two to ninety per cent, except those which are principally seed."

"Do you consider the banana a wholesome food?"

"It does not agree with most people."

"Why is that?"

"That is because it is pulled green, and ripened by an artificial process, so that when the banana is ripened for market, it is really ripened by a process of decay."

"Then this is the reason why bananas are so likely to disagree?"

"Yes, being partly decayed, and containing a considerable amount of sugar, they are likely to continue to decay, or sour fermentation set in after they are eaten. It is no uncommon thing for bananas to produce nettle-rash, especially in children."

"Is there any way of overcoming the difficulty?"

"Only by allowing the banana to ripen where it grows, and make it into meal. This is another peculiarity of the banana; it is the only fruit that can be dried and ground into flour, and when this is done the banana makes a valuable food."

"I notice that its use has been mentioned in typhoid fevers?"

"Yes, banana meal has been used with very good results in many hospitals, both for typhoid fever and other cases, but it must not be concluded from this that an ordinary tough banana can be used, because it would likely disagree with a well person, and be very dangerous to the sick."

"Then you rather discourage the use of bananas?"

"Yes, until there is some way of getting the fruit to us
in a better condition.   It is truly a fine fruit and the time
will soon come when its use will be such as to warrant
some more satisfactory way of bringing it to the people."

"I suppose, Doctor, that you consider the lemon more
of a medicine than a fruit?"

"The lemon has long been used for its flavor, and in a
medicinal way, but modern chemistry so perfectly coun-
terfeits all flavors that the use of fruits for such purposes
is almost discontinued, so that the lemon must hold its
place for its valuable acids."

"What are these?"

"Citric and malic acids.   A lemon does not contain any
properties that could really be called a food, and its use
is really only that of a cleanser.   It is especially valuable
to cleanse the stomach of mucus, when its juice is used
with hot water an hour before meals.   No sugar should
be used.   As a toilet article for the skin, hair, and mouth,
it has no equal, for its juice cleanses the skin of an excess
of fat, and should be used to take away the "shine" on the
face, the ladies so much dread, instead of face powders.
It will also remove blackheads, due to impaired circulation of the skin, and is truly nature's beautifier.   The juice
of the lemon when used without soap is an invaluable hair
wash to remove dandruff and oil, and will also cleanse and
sweeten the mouth, when there is a bad, or 'dark brown'
taste.   For washing the hair, the juice of a fresh ripe
lemon should be squeezed into a pint and a half or two
pints of lukewarm water, and thoroughly rubbed into the
scalp, then dried with a rough towel."

"Has the orange the same uses?"

"No, it is more of a food, because the orange contains
a little gum and some sugar."

"What acid does the orange contain?"

"Mostly citric and malic acid, and citrate of lime."

"Then the orange has uses unknown to the lemon?"

"Yes, oranges are often valuable for invalids when lemons could hardly be used at all."

"The tamarind has a high per cent of citric acid, also contains some tartaric acid, and a trace of malic acid. It is rich also in potash, and contains as high as 12% of sugar. It is not extensively used, and does not therefore deserve much consideration."

"The pineapple is one of the most delicious of all tropical fruits. It contains all of the fruit acids and some other substance very similar to papain, which is a digestive agent for all kinds of food. There is probably no other fruit generally known that has the same property for digesting other foods as that of the pineapple."

"Then it is a good thing to eat?"

"That is very questionable. The pineapple contains an extraordinary amount of tough fiber, which is exceedingly difficult to digest.

"Then how should it be used?"

"Well, the juice should be obtained in some way from the pineapple without the tough fibre, macerated in water and expressed by compression. It is now prescribed to considerable extent in certain diseases of the stomach. Zumo-Anana is a pineapple wine, beneficial when there is insufficient secretion of digestive juices, but contra-indicated where there is excessive secretion.

"The lime is probably the sourest fruit known. Citric acid is manufactured from it. Also, lime juice. It is very similar to the lemon. Citric acid is often used as a substitute for lemons.

"The grape fruit is a large fruit, much larger than either orange or the lemon. It contains similar properties to the lime and lemon with some bitter matter. It is not extensively used, but makes a cool and refreshing drink, and a few people like the fruit."

"Are there no important fruits, other than what you have discussed?"

"Yes, dates and figs. The dried dates and figs of commerce are the richest of all the fruits."

"What is the average composition of them?"

"They contain more than two-thirds solid matter, about four per cent of flesh-forming substances; and nearly fifty per cent sugar, considerable waste material, and mineral matter. The date contains very nearly the same properties as the fig, with the addition of pectose or gum. These fruits dried contain nearly the same proportion of heat-producing and tissue-forming substances as rice, and will therefore support life for a considerable length of time."

"Are they used extensively as food?"

"Not so extensively as they should be, for figs are quite laxative, which is due, partly, to the seeds, and partly to the fact that figs, especially green figs, have a digestive agent similar to that of a pineapple, only less pronounced. They are used more by vegetarians than others, and deserve a favorable place in our dietaries, but should always be cooked."

"Olives are only used for two purposes in this country; that is, we use the oil made from the olive and the green olive for pickles. The oil is valuable, but the pickles are tough, and have no use as food. Large doses of olive oil have been recommended for the removal of gall stones."

"Citrons have no other use except for flavoring, but not many people like them for that purpose. They are tough and well-nigh insoluble, and should not be eaten for food."

"Doctor, I believe it would be good for you to suggest something about preserving fruits."

"Fruits are usually kept either by being dried or canned. Drying is an easy process with proper appliances. Sun-dried fruits are better than no fruit at all; but any slow process of drying where the fruit is exposed to the atmosphere, furnishes the best opportunity for all kinds of insects and bacteria to secure a lodging place, so that sun-drying, or any slow process of drying, should be avoided as far as possible. Fruits that are quickly dried in closed ovens are very much better, and if packed at once and kept from exposure they will be much less likely to be infested with insects, and in every way superior to sun-dried fruit."

"How should fruit be prepared for both canning and drying?"

"The peel, core and all damaged places should be removed. It is a very bad practice to either dry or can fruit with the skins, for they cannot afterwards be removed, and the skin is particularly objectionable in dried fruit, and any other kind, unless thoroughly cleansed before being canned."

"Why is it that so many people do not succeed in properly preserving canned fruit?"

"Because it is not properly canned."

"What is wrong with the ordinary method?"

"The principle of canning fuit merely involves the destruction of bacteria, and then closing the cans so that neither they nor air can enter it."

"How should this be done?"

"It is best to cook the fruits in the cans, so that no bacteria can enter in filling them. If this cannot be done, the cans should be set in hot water after they are filled. Probably where more failures are made than anywhere else, is with the lids. It is not only necessary to have the cans thoroughly sterilized by being boiled in water, but the lids must also be sterilized. If fruit cannot be cooked in the cans, the lids should be sterilized and put on the cans with a small vent for escaping steam. If they are then immediately sealed, so that they are air-tight, there will be no trouble in properly preserving them. It must be borne in mind that nothing should touch spoons, lids or anything that comes in contact with the fruit after being sterilized in boiling water."

"What do you mean by sterile or sterilizing?"

"Anything is said to be sterile when it has been subjected to a degree of heat sufficient to kill all kinds of bacteria."

"How much heat is ordinarily required?"

"There are very few microbes of any kind but what are killed after being subjected to boiling water for, say, fifteen minutes. A high degree of heat, if it be moist, such as steam, answers the same purpose, or better. The whole theory of preserving canned goods rests upon the destruction of bacteria and the elimination of air; and as the microbes cling to every known substance, it is necessary to have the hands perfectly clean and all the instruments or vessels sterilized in which the fruit is handled, as already suggested. It is best to cook the fruit in the cans with the lids on. This can be done by filling the cans and setting them in a kettle of boiling water, so that the cans are almost entirely covered. This prevents the entrance of bacteria from handling, and sterilizes the fruit in the jar."

"Which do you consider the most important of the nuts that are used in this country?"

"The peanut. The consumption of peanuts has grown

to be enormous, and is destined to be many times greater than it is."

"Why do you say that?"

"Because the peanut is a palatable and rich food, and it supplies most of the necessary elements to sustain life."

"What properties has the peanut?"

"The largest ingredient of the peanut is its oil, amounting to about fifty per cent. It has, in addition, considerable gum, the equivalent of starch. The mineral mat-cent. waste material."
ter amounts to nearly two per cent., and about four per

"What about the tissue-forming substance?"

"The nitrogenous part of the peanut is high, amounting to twenty-four per cent or more."

"But, Doctor, it is said to be a great source of dyspepsia."

"It is at least fair to say that it is very difficult to digest."

"Why is this?"

"Because it is really a concentrated food; practically, it has no water, and consequently it is exceedingly solid. It naturally follows that the digestive juices will not penetrate the particles very quickly. Very few persons will masticate the peanut to finer particles than cracked wheat."

"How can this difficulty be overcome?"

"Only by grinding. Extraordinary care in masticating peanuts by keeping them in the mouth as long as possible, overcomes part of their objectionable texture."

"Why do you think the peanut has a great future?"

"Because nearly everybody likes it, and it supplies nearly everything necessary to live on, and is comparatively cheap. It is only a question of time until it is better prepared and furnished to us so that it can be used with other foods; for it seems admirably adapted to furnish both the necessary oil and flavor for the cereals, which are deficient in both."

"What about other nuts?"

"All nuts contain a large per cent of oil. The chestnut is the only one that contains a great amount of starch; probably the hickory nut is really the most palatable of all, and is rich in oil."

"Are not pecans good?"

"The pecan has a bitter shell which makes it disagreeable, if any particle be left in contact with it. It does not vary greatly in composition from the hickory nut. The only nut having special use is the almond."

"What special use has it?"

"As it does not contain any starch, has an agreeable flavor, and is quite a rich food, both in tissue-forming and heat-producing substance, it is very valuable in Bright's disease. It is exceedingly tough and solid, but probably not so much so as filberts and hazelnuts. These are also rich nuts, but need grinding more than any of the others, and unless they are ground, they are exceedingly indigestible."

"Among nuts, what prominence would you give walnuts?"

"The black walnut is rather a strong-flavored and very oily nut. The white walnut, or butternut, is still stronger in its flavor, but not so rich in oil. The English walnut, so-called, which is principally grown for market, is a rich, oily nut. It is not so firm as many of the other nuts, and has some advantages over them. Probably the most oily of all nuts is the Brazil nut. They are also quite firm, but almost pure oil."

"Are there no other nuts that you think worthy of notice?"

"None, unless we except the cocoanut, which is becoming quite an article of commerce, especially its oil. It is now used extensively for making soap, and other purposes."

"Do you consider the cocoanut a good article of food?"

"I do not. It is one of the toughest and most indigestible of all articles used for food; even shredded cocoanut is extremely difficult to digest,.and the only way that it can be ever used successfully as a food, is to provide some way of pulverizing it to make it as fine as flour, or nearly so, which would not only make it digestible, but more palatable, as well."

"Then, Doctor, you are quite a friend to nuts, provided they can be used properly; but your declaration that they are wholesome don't agree with common experience."

"Well, if the people would undertake to live on any of the cereals without their being ground or cooked, the results would be worse than the ill effects commonly attributed to nuts. The Sanitas Food Co., of Battle Creek, Mich., are making nut foods that are as much easier digested, compared with raw nuts, as Granose or Granola, compared with unground and uncooked wheat. Nuts have long been known as rich food, but owing to their solid texture, and the natural inclination to swallow them in uncrushed particles, they have, for many people, been considered rather indigestible. The Sanitas Food Co. have overcome this difficulty and given the world. the most delicious and fattening foods ever manufactured. They answer every purpose of meat, and greatly strengthen the cause of vegetarianism."

"Doctor, if you connect anything with vegetarianism it will prejudice it in the estimation of some people."

"That ought not to be so. Much of the prejudice

against vegetarianism is due to the fact that most vegetables do not supply either enough fat or tissue-food.
These defects are supplied by using the entire grain of
such cereals as wheat in connection with nuts, as they are
rich, both in fat and tissue-forming elements."

"Then nut fats are superior to animal fats?"

"Yes. Heretofore cream has held first place among
common fats, but the nut-cream and nut-butter, made by
Sanitas Food Co., are superior to either cream, butter
or animal fats."

"In what particular?"

"All animals are subject to disease—cows especially
to tuberculosis—besides, cows are frequently kept in foul
places, milked by soiled hands, and the milk kept in unsanitary places and in vessels washed in water containing
typhoid or other bacteria. These dangers are avoided
in the nut foods; but there are still stronger reasons for
their use. The particles of fat are so minutely subdivided
or emulsified, that they are readily taken up in the system. The animal fats will not sustain life, as they contain practically nothing but heat-producing elements. The
nut foods will sustain life and more quickly fatten than
anything yet discovered. Nut butter and almond butter
are the most delicious and appetizing fats ever produced,
and they will very likely displace cod liver oil as a fat-
producing food for consumptives."

"Why not combine nuts with grain foods?"

"The Sanitas Food Co. has done so with great success.
Long ago I was impressed with the belief that emulsified
nuts could be combined with dextrinized or pre-digested
starch so as to make the richest and best food for fattening yet discovered. The Sanitas Food Co. has made
such a food and named it Bromose. As a fat-producer
and food-tonic, Bromose has produced most remarkable

results.   Nuttose is another similar food, and might aptly
be called vegetable meat.   Granose and Bromose used
together have restored many invalids to vigorous health.
Knowledge of these foods are of so much benefit that I
have spoken of them at some length."

CHAPTER XIX.

CONDIMENTS AND DRINKS.

"Doctor, what do you mean by condiments?"

"Well, the word has a general and well-understood meaning, but, for my use, I would explain it by saying that it should be pronounced by emphasizing the second syllable and sounding the 'i' long, making it con-die-ments."

"Then how would you define it?"

" I would say that it is the thing we eat with our food

"But it is a hot subject."

"That is a hot criticism, Doctor."

which beguiles us to death."

"You mean by that, that condiments burn?"

"Yes; take pepper; it irritates and burns the membranes very much like fire."

"But the doctors say it aids digestion."

"So it would warm your hand to put it in the fire."

"But that would destroy the hand."

"So the pepper has a tendency to destroy the digestive organs."

"If that be true, how can it aid digestion?"

"Anything that irritates the mucous membranes of the stomach increases the gastric secretion, and this is what pepper does; but in doing this, it inflames the stomach, causes excessive secretion of acid, and an uncertain number of stomach disorders."

"Is that why so many people want to drink ice water at meal times?"

"It is **one** of the reasons. If a mouthful of pepper be swallowed, the first impulse is to drink some cold water as quickly as possible to relieve the burning sensation.

"When the stomach is habitually irritated with pepper, mustard, alcoholic liquors, horse radish, or indigestible food, cold drinks give a feeling of relief, but really only aggravate the irritation?"

"Then pepper is injurious?"

"It is most injurious, and should have no place in our dietaries, and should only be used as a drug."

"Is pepper worse than other condiments?"

"Probably not; mustard, sage leaves and horse radish are all bad, the two latter being worse, if anything, than pepper."

"How about salt?"

"Salt is much railed at by a certain class of hygienists, but is strongly defended by others."

"Which side is right?"

"If we are allowed to make comparison with the lower animals, it would seem that carniverous (flesh-eating animals) care nothing for salt, while herbivorous (grass-eating animals) in all countries are intensely fond of salt."

"That would indicate that man, living largely on the vegetable foods, would require some salt?"

"Yes; but not more than a quarter, and probably not more than one-tenth of what most people consume. Excessive salt eating is a bad habit, but not so bad as the use of pepper. It has been demonstrated that salt retards digestion, causes skin eruptions and other derangements of the system; and, while salt is useful, most persons would be benefited if they would cut down their salt eating to one-fourth of what they are in the habit of using."

"Does vinegar belong to the same class?"

"Vinegar is bad, but not wholly so; for it has some uses."

"Haven't you already said that vinegar must be discarded?"

"Yes; its uses are for the most part injurious, but still it has some use, when better acids are not obtainable."

"Then why abandon it?"

"The abundance of fruits we have furnish us acids so much superior to that of vinegar, the question of continuing the use of vinegar ought not to be considered at all, when fruits can be obtained."

"Then you would use sour fruit on salads, instead of vinegar?"

"Yes, or else not eat the salads at all."

"Why is vinegar so objectionable?"

"Because it is a ferment filled with vinegar worms, which can be seen with the naked eye. If you will get some vinegar plant and examine it, you will not care for the vinegar afterwards."

"What fruits would you recommend instead of vinegar?"

"Limes, lemons, grape fruit, sour oranges, sour grapes, sour berries or even rhubarb."

"How about spices?"

"Some of the spices are not objectionable."

"Which would you give first place?"

"I would give the first place to nutmeg."

"Why so?"

"Because it has an agreeable flavor, is mostly oil, and is not particularly objectionable in any way."

"You favor nutmeg, but condemn other spices?"

"All-spice is not very injurious, but its flavor is not pleasant. Few people care for all-spice as a flavor, but almost everybody likes cinnamon."

"Then you would strongly recommend cinnamon?"

"Well, I would not recommend cinnamon or peppermint for uses in food except as a medicine."

"Why so?"

"Because both cinnamon and peppermint kill bacteria to a certain extent, and are known as antiseptics. They would both have a tendency to arrest the processes of digestion. They also have a tendency to arrest processes of decay, and both are useful to relieve a sour stomach. Owing to this fact, they are not desirable to mix with food for general use, but are valuable for special uses, where their antiseptic properties are needed."

"How would you class cloves, caraway seed, ginger, etc.?"

"Cloves are very astringent, but the small quantity used is not likely to do harm. Caraway seeds cause nausea in many people, which shows that they are irritating in their nature, and should not be used at all. Ginger is a very pleasant stimulating condiment. It is quite useful, too, when such stimulants are needed, but it is not desirable for habitual use."

"Doctor, you are much against the ordinary spices used by the cooks, what about tomato catsup, pickles, etc.?"

"As to tomato catsup, if it does not contain irritating substances other than the tomato, its use is not particularly objectionable, except where acids are harmful; but as to pickled cucumbers, onions, cabbage and olives, they are all tough, and if one cares to be free from aches and pains, he should let all of them alone."

"You have not mentioned olive oil, Doctor?"

"There is no objection to olive oil, and it may, at times, be very useful, where such food is needed."

"Are these all the spices or condiments in general use?"

"All the important ones, although anise, fennel, parsley, sorrel, are used to some extent, but not enough to be of any particular injury or benefit either."

"Then you don't advise the use of flavoring or spices at all?"

"There is one general objection to all of them, and that is this: they have a tendency to stimulate the appetite and cause one to eat too much, and as over-eating is a practice well-nigh universal and injurious, the things that favor it should not be encouraged."

"Do I understand from this that you would not use flavorings at all?"

"I did not mean that, but they should be used to make those foods we do not like, but ought to eat, more palatable."

"Give us an illustration."

"A great many people suffer from uric acid headaches, the result of meat diet and constipation. Now, such persons may not care for cereals or coarse vegetables, and would, therefore, not eat them, to any considerable extent, because they do not like them. The proper use, then, for flavors, is to take the foods that such a person ought to eat, and flavor them so they would be agreeable. This would make the coarse cereals palatable, and if substituted for their meat diet, their headaches would disappear, and their health be entirely restored."

"Doctor, suppose we go and take something?"

"All right; I will drink a glass of mineral water and then discuss drinks."

"Doctor, since you do not drink anything very stimulating, you probably have something caustic to say about liquor drinking?"

"I don't propose to commence on alcoholic liquors, but on the drinks that pave the way for them."

"That is a new idea. You don't mean to say that other drinks cause an appetite for liquor, do you?"

"Such a thing as an appetite for liquor, strictly speaking, does not often exist."

"What, then, is it?"

"It is a mental condition which makes the individual crave stimulants."

"Then it is a craving for the effects and not the taste?"

"That is it. A well person is free from nervousness, and does not want any stimulants."

"Then our habits affect the nervous system?"

"Yes; nervousness apparently increases with each generation. It is often attributed to worry, but the real cause is the habits of the people, and a large share of it is due to their drinking habits."

"Then nervousness does not result merely from worry or overwork, as many people suppose?"

"No; people are mainly worried because they are nervous. If we were not nervous, the ordinary cares of life would not cause us to worry."

"Doctor, I am very anxious to know what you are going to charge this to?"

"To no one thing; but I wish to show the relation of other causes to nervousness."

"Since you have already spoken of water, I suppose you will charge tea and coffee with a good deal?"

"Yes, coffee in this country, and tea in other countries."

"I thought the general opinion was that coffee was not injurious?"

"General opinion is about as safe to guide us as it would be to have a mule put in a pilot house to steer a ship across the ocean. As an illustration, there are numbers of people who are sick every week, or at least every month, and are foolish enough to say that nothing they eat or do, hurts them; and it has often struck me that it would be just as reasonable to say of a man who is hung till he is dead, that the hanging didn't injure him, but that he merely died because he stopped breathing."

"Why do you say such things?"

"To get people to understand that when they are knocked over, something struck them."

"What bearing has this on coffee?"

"That effects have causes."

"How can you prove that coffee has any bad effects?"

"By drinking it or watching others."

"That doesn't throw any light on the subject."

"Well, coffee is a stimulant, and the heart has only a limited capacity. When it is stimulated beyond that, it must be correspondingly weak, just as it was stimulated to increased activity by the coffee. Suppose we illustrate it in this way: We will take two tanks of water and connect them with a pipe. Now, if each tank be two-thirds full, the pressure will be equal; but if the water be pumped from one to the other, the one will have its pressure increased just to the extent that the water is taken from the other, and the one from which the water is taken will have its pressure decreased."

"Now are you sure that the effects of coffee are similar?"

"It is very much like it; people who drink two or three cups of strong coffee could hardly get along without it. If they do not have it, they will have the headache and be irritable, or, in other words, there is relaxation, a weakness of circulation and the machinery of the system refuses to run properly until it is again brought under the influence of a new supply of coffee."

"Well, I suppose I have seen hundreds of people of that kind, but I never thought it was serious."

"Yes, it is; if coffee is necessary to keep any one going, so to speak, such person might properly be called a coffee inebriate."

"Well, how does that pave the way for liquor?"

"In this way: People who are affected by the use of coffee become nervous to an extent that is chronic, and the condition of the nervous system of the parents is likely to be transmitted to their children."

"Then it is not so much of an appetite as it is a nervous condition?"

"No; liquor drinkers, or at least very few of them, will admit that they drink liquor because they like it. It is purely an abnormal craving for some kind of stimulant. They don't feel right without it."

"Then why don't everybody take to the use of strong drinks?"

"A very large per cent. do, but it must be remembered that not every one uses stimulants to make a serious nervous condition. Besides this, nature constantly attempts to correct her own defects, and if it were not for the fact that each generation keeps imposing upon nature, we would soon be an ideal race."

"Has coffee no uses?"

"Yes, coffee is useful as a drug, is very pleasant to the taste, and if water and milk are merely flavored with coffee, and it is used as a flavor more than something to tone up the nervous system, it is not seriously harmful, and may be, at times, very useful. It may not injure every one, but the habit of drinking two or three cups of coffee (strong coffee at that) is most pernicious, and not only does serious harm, but is very likely to do harm to unborn generations, as well as to lead to the use of stronger stimulants."

"Are there any other ill effects from coffee?"

"A considerable quantity of strong coffee retards digestion and does injury in that way."

"What property is it in coffee that affects the nervous system?"

"The name given it is caffeine. It contains some other properties, but it is the caffeine that gives it its stimulating effects. The practice of giving coffee to children cannot be too strongly condemned."

"What are the symptoms of coffee inebriety?"

"Nervous tremor, languor, prostration, sleeplessness, craving, with headaches when it is not supplied in sufficient strength or amount."

"If coffee be used at all, how should it be made?"

"Coffee is the least harmful when the smallest amount of its active properties are extracted. The longer it is steeped in water and the more it is boiled, the more injurious it is. It should not be strong, and should never be permitted to boil; nor should it be permitted to stand for great length of time, but the hot water should be poured on and left only for a few minutes, to merely extract the aroma of the coffee."

"Doctor, how does tea compare with coffee?"

"It is very similar to coffee, and there is much discussion as to which is the more wholesome or harmful."

"What is your opinion?"

"Tea, in my opinion, is still worse than coffee. It is perhaps not quite so much of a stimulant, but contains a high per cent. of tannic acid. This makes tea an astringent and a great source of constipation. It therefore deranges the system, causes nervousness, insomnia, and has more serious effect on the digestive organs than coffee."

"In what way does it affect digestion?"

"Well, it affects digestion directly, because of the fact

that the tannic acid of tea precipitates albumen in the foods. This may be better understood by saying that if tannic acid be applied to the white of an egg, which is albumen, it will make a tough substance of it quite like leather. It is, in fact, this process of applying tannic acids to skins that makes leather."

"Is that the reason why, that when egg is used to clarify coffee it forms in lumps?"

"Yes; that is the same principle. It clarifies the coffee by being coagulated and gathering the coffee grounds as it settles, and when egg is used to clarify coffee or any other drink it should be strained."

"Doctor, you say tea and coffee are both bad. They are certainly not worse than tobacco?"

"No; they are not worse than tobacco; but tobacco does not belong to foods, and we cannot discuss it now."

"Is there any other objection to tea and coffee?"

"Yes; both tea and coffee are especially injurious to persons who have what the physicians call the arthritic tendency."

"What do you mean by that?"

"A large number of diseases result because the system does not perfectly clear itself of nitrogenous matter. These various ailments are styled uric acid diseases."

"What are some of them?"

"Probably the most universal one is sick headache. Periodical sick headaches are almost certain to be the result of uric acid in the blood."

"Then tea and coffee add to the uric acid in the system?"

"Yes; people who have sick headache, asthma, bronchitis, rheumatism, gout, epilepsy, and diseases of the stomach generally, should drink neither tea nor coffee."

"Has tea any particular use?"

"It is sometimes used because of its astringent properties in diarrhoea, or summer complaint, and is really more of a medicine than food."

"Doctor, from what you say, I conclude that the abuse of tea and coffee is almost as far-reaching in its effects as that of alcoholic liquors?"

"You are right; and it is really a pitiable sight to see people who are trying to save the world from its vices adopt such habits that they transmit such traits and characteristics to their own children, that the very evils they seek to remedy are fostered by their own offspring."

"Cocoa is a preparation made from the bean or seeds of the cacao tree."

"What is its composition?"

"It contains a very large per cent. of oil, theobromine, which is very similar to the caffeine in coffee, and theine in tea. The cocoa also contains a little albumen, starch, fiber and mineral matter."

"Where does cocoa come from?"

"Principally from the West Indies, Brazil and the northern countries of South America."

"How is cocoa prepared?"

"The seeds grow in a pod and are removed, dried, fermented and then ground. Each manufacturer of cocoa, of course, having peculiar ways of preparing his product."

"How are the ordinary preparations made?"

"After the seeds have been treated as described, they are made into a paste, to which starch and sugar is added."

"Do you consider it a good drink?"

"It is less injurious than tea or coffee, and owing to the amount of oil it contains, it is much richer than either of those. People who do not tolerate fats or oils, should not drink cocoa. It is slightly stimulating, and contains some insoluble matter. Some people do not like cocoa, because it presents an unsightly appearance, on account of a scum of oil appearing on top of the cup. If this is not relished, it may be skimmed off, either with bread or in some other way. Cocoa butter is said to be a very agreeable and useful oil."

"For what purpose?"

"Many people prefer cocoa butter to any other fat or oil for table use. It melts at a very low temperature, and is very easily dissolved. It is also used to a very considerable extent for the administration of drugs in capsules or suppositories, and for anointing the skin in eruptive fevers; also useful as inunction while massaging."

"Chocolate is made from the same material as cocoa, but is deprived of part of the oil (it is supposed that chocolate contains some of the husks, as well as the seeds). It is not quite so rich, ordinarily, as cocoa, and some people prefer it. Chocolate is also used extensively in confectionery. It might very well be used to flavor foods to make them more palatable. It is also a nutritious food when taken alone."

"Doctor, you don't seem to favor chocolate for general use; you say that tea and coffee are both very harmful, and that many who use them are really tea and coffee inebriates; now, recognizing the universal desire for some drink, what can they drink that will not harm them?"

"Cereal coffee. The Sanitarium Food Co. make what they call Caramel Cereal. It is a pleasant and harmless drink, and will greatly benefit the nervous, anaemic and dyspeptic. It has no particular food value, and its manufacturers claim none. The benefits of any cereal coffee are purely negative, and in that lies their value; they do not poison; coffee does. Claims for great nourishment from cereal coffees cannot be sustained."

"Doctor, you stated awhile ago that tea and coffee were

abused to such an extent that their effects were almost as far-reaching as alcohol."

"Yes; but I did not mean by that that the effects were as violent."

"What is ordinarily understood by the term alcohol?"

"Used in a general way, alcohol includes all alcoholic liquors, such as whisky, beer, wine, ale, gin, as well as chemically pure alcohol."

"Is there not much disagreement as to the value of alcohol when moderately used, as to whether it is really a food or not?"

"It has been demonstrated that the system will absorb a small amount of alcohol because it is not exhaled in the breath and cannot be found in the excretions. It is, therefore, fair to conclude that it is burned up in the system, just the same as fats and starches are."

"Then it would seem from this, Doctor, that if a small amount of alcohol is absorbed as food, it would not be injurious?"

"But you forget that I have already explained that a healthy person has no need for stimulants, but that they are an injury instead of a benefit. This is especially true of alcohol."

"Why is it more true of alcohol than any other stimulant?"

"Because it causes the tissues of the body to be changed, although it is claimed that some persons can use a small amount of alcohol without any discoverable change in the tissues of the system."

"Then, if that be true, it must explain the constant tendency of those who use alcohol to increase the quantity?"

"Yes; the stimulating effect is what is always sought, and as the change takes place in the tissues, it takes more

and more to have the same effect that the small amount
originally had." .

"If this be true, it is a dangerous thing to use, purely
upon physiological grounds?"

"Yes, it is dangerous to make a practice of using any
kind of alcoholic liquors, for the reasons explained."

"What is the character of the changes usually made by
the continued use of alcoholic liquors?"

"It changes the texture of the liver, blood vessels,
kidneys and the digestive organs generally. It makes
the individual much more susceptible to infectious diseases, especially typhoid fever and pneumonia."

"Are there any other ill effects resulting from the use
of alcoholic liquors?"

"Yes: all those cases described under arthritic tendencies resulting from uric acid in the blood, are unfavorably
affected by all alcoholic liquors, or, more especially, malt
liquors, that contain acids, as nearly all the malt liquors
and wines do, so that persons having sick headache, rheumatism or kindred ailments, must not use liquors at all."

"Doctor, doesn't the use of alcohol lessen the desire
for food?"

"Yes, if a certain amount of alcoholic liquors is used
up by the system, it corresponds or equals a considerable
amount of heat-producing food; and as it contains no
waste material, it would naturally have a tendency to produce constipation, which is a well-known fact."

"Are these all the objections to the use of alcohol?"

"Not all of them. There is still another. That is, that
alcoholic liquors lower the temperature of the system."

"I thought it produced heat and people drink liquor
in the winter time to warm them?"

"In cold climates they know better. No alcoholic

liquors dare be permitted in a northern lumber camp."

"But it makes a person feel warm."

"That is true. It brings the blood to the surface, and for that reason assists in the radiation of heat; or, in other words, it causes the blood to flow out in increased quantities, so that it can cool faster, and it has been demonstrated, hundreds of times, that one can stand much more cold without the use of liquor than with it."

"Then, if that be true, the use of liquor ought to reduce fevers."

"It is often given for that purpose with good results."

"Then, from your point of view, alcohol is a medicine, and has no place as a food?"

"That is true; the habit of using liquors for the purpose of making one feel better results in untold harm. People ought to learn how to live so that they do not need liquor or coffee or any other stimulant to give them energy sufficient for their work."

"What is the principal effect of alcohol as a medicine?"

"It stimulates the nerves, the action of the heart, and increases the circulation. It is sometimes useful in diseases of the stomach as a stimulant and to prevent decay, but should never be used except by the advice of a competent physician."

"Then, according to your view, alcohol does not give increased strength, as many people suppose?"

"No; it merely acts as a spur or as a whip would on a tired horse, and only draws on reserve force; and the more that is drawn on, the sooner it results in total incapacity for work.

"Probably the most important of all the uses of alcoholic liquors is in fevers. When, after prolonged illness, there may be danger of the heart's failure, liquors are useful in such conditions to stimulate the action of the heart."

"Doctor, it is a common belief that the use of beer decreases the consumption of other liquors?"

"That is very doubtful. In the first place, it is difficult to get reliable statistics, because it is almost impossible to determine the number of people who drink liquor. It is a well-known fact that the total abstainers have increased more rapidly in recent years than ever before; and if this be taken into consideration it is more than probable that the liquor drinkers still drink as much or more distilled liquors than ever before, notwithstanding the enormous consumption of beer, at least the general use of malt liquors has increased, rather than decreased, drunkenness."

"Doctor, you have said nothing about the moral aspect of liquor drinking."

"Well, we are not dealing with moral questions, but we have already said that most people have neither the instinct of brutes nor sufficient reason to guide them, and are, therefore, very imperfectly organized."

"When can liquors be drank with the least injury?"

"At meal time; for some people, a small amount of liquor at meal time is less harmful than tea or coffee; but the man who drinks quantities of liquor of any kind whatsoever, on an empty stomach, so violently outrages his system that he needs something added to or taken from his brains."

"Doctor, what kind of a drink is soda water?"

"Soda water is made by simply aerating water with common carbonic acid gas."

"How is the gas made?"

"By saturating marble dust with sulphuric acid, and the gas is collected in a tin-lined tank, and is drawn off, mixed with the water as drawn. The flavoring matter is supposed to be fruit syrups; but for the most part, they are made of essential oils."

"Is the drink beneficial or injurious?"

"There is not enough of any very active substance to have much effect, and it is usually considered an occasional glass of soda water does no harm, although if tanks are not properly lined, the soda water would be poisonous. It should not be drunk except on an empty stomach."

"Ginger ale is quite similar to soda water, only has considerable quantity of ginger. It is sometimes beneficial, but not a good drink for continual use."

"Root beers are usually fermented drinks; and while some people think them wholesome, it is difficult to understand how any fermented drink can be of any use to the system, although they have probably not sufficient amount of ferment to be of any particular injury, and certainly no benefit.

CHAPTER XX.

INFANT FEEDING.

"Doctor, you have discussed the processes, of digestion and the properties of foods, cannot the people adapt their diet to their needs from the information you have given?"

"Well, the advice ought to be of great help, yet there is more to learn; for we have not said a word about the quantities of food needed for different conditions."

"A good many people think that appetite ought to govern, both in the selection and amount of food that each person should eat."

"What else has been governing them since the dawn of creation? As the people now live, ninety-nine out of every hundred are partly or wholly disabled a considerable portion of the time, which is a poor reason for doing things as they have in the past. Those who say that appetite should absolutely govern, are not very thoughtful, to say the least."

"Why so? Do you think present conditions show such grievous effects?"

"Not the effects alone, however bad, but the principle of being governed by appetite is not in harmony with common practice; for is it not the cook who decides what food we shall eat?"

"I guess you are right, Doctor; the cook may not prepare one meal a week to suit the appetite or needs of a single member of the family."

"That's the point; most people have little or no choice in the selection of their diet; they must eat what is furnished, and that is often incompatible; hence, the system is not nourished, and an excessive amount of food must be consumed to supply some necessary ingredients. It is foolish to talk about instinct guiding a human being. If such were possible, it would be the least perverted, and therefore a safe guide in the care of infants. Now, a child will go into the fire, or off of a precipice, or swallow pins, coins, buttons, and often kill themselves eating pop-corn, raisins and other foods; while an animal, governed by instinct, will not do such things. Here is another very striking illustration of both ignorance and lack of instinct. Young infants often have indigestion from nurs-ing too frequently. This gives them the colic, so-called, and their discomfort makes them fretful."

"Well, what of that?"

"The baby cries, and the mother hastens to nurse it. Now, it is already suffering because it has nursed too often, but the child has no instinct and the mother no knowledge to prevent the repetition of the injury."

"Your statement seems reasonable, Doctor; for almost every one knows that a grown person cannot stand con-tinual feeding, and it does not seem rational to conclude that a young babe could do it."

"No; the injurious effects of continual feeding have been so often proven by every good physician, it must be accepted as a fact."

"Doctor, how often should a baby nurse?"

"During the first three days after birth, four or five times a day. One or two teaspoonfuls of water may be given occasionaly, but no other food. After the first few days the child may be allowed to nurse every two or three hours, between 5 a. m. and 11 p. m., and once during the night, until five or six months old."

"Should the child be fed at regular intervals?"

"Yes; it is of greatest importance. The hours for feeding should even be more regular than that of a grown person."

"Doctor, would it not be a good idea to give a table, showing how children should be fed at different ages?"

"Perhaps so; the best authorities give the following as a guide to hours and quantity of food required for a child up to one year of age:

Age of Child.	How often fed or nursed.	Number of times fed during night.	Amount of each feeding.	Daily total.	Daily number of feedings.
1 week	2 hours	2	1 oz.	10 oz.	10
1 weeks	2 "	2	1½ to 3	15 to 16	8 to 10
1 month	2½ "	1	2¼ to 3	20 to 24	9 to 10
2 "	2½ "	1	4 oz.	28 oz.	7
3 "	2½ "	1	4 oz.	28 oz.	7
4 "	3 "		5 oz.	30 oz.	6
5 "	3 "		3½ oz.	33 oz.	6
6 "	3 "		6 oz.	36 oz.	6
9 "	3 "		7½ oz.	37¼ oz.	5
12 "	3 "		8 oz.	40 oz.	5

Of course the size and vigor of the child make it necessary to vary the quantity accordingly."

"Will there not be a tendency to fretfulness between the periods of nursing?"

"There should not be, although babies frequently get dry and cry for water. If a child is fretful and there is no reason why it should be hungry, it should be given water with a teaspoon."

"How soon can the night nursing be discontinued?"

"After a child is six months old it may be nursed at bed time, say ten o'clock, and early in the morning, before seven. After it is a year old, it need not be fed later than seven or eight in the evening and at its usual hour of awaking in the morning."

"Will babies readily accept this arrangement?"

"Not always. They may want to nurse every fifteen minutes; but the mother should be guided by reason in feeding, just as she would in keeping her babe out of the fire if it should have an impulse to go into it."

"Suppose the child doesn't thrive, what then?"

"Of course no arrangement of feeding can supply the place of wholesome milk, and it often happens that the mother is incapable of doing this. In such cases the next best thing is cow's milk."

"That would seem to be very poor, considering the number of deaths attributed to it."

"Artificial feeding has always been the greatest source of infant mortality, and great care should always be exercised in the preparation of milk for infants. Cow's milk differs greatly from human milk."

"Yes; I remember that you said it contained much more casein, or curd, and much less milk sugar."

"So I did; and it is therefore much more difficult to digest, and should be modified for infant feeding."

"What do you mean by modified?"

"It must be diluted to make the curd smaller, and enriched by cream and sugar."

"What is the best method of doing this?"

"It should receive about twice as much water as the quantity of milk, so that one pint of milk makes three after being diluted. This may be done in several ways. If the child's digestion be good, pure water may be all that is required; but if not, and the child is sick or cross, some other method must be resorted to. The most common diluent is barley water. For this, take pearl barley (or rice) and pound or grind to a fine flour; add two tablespoonfuls of the flour to each quart of cold water and boil for an hour, and then strain through clean, fine linen or a colander. Keep in cool place. In case of diarrhoea, lime water will be most useful. Take a lump of unslaked lime, half the size of an egg, and pour two quarts of hot water on it, and let it stand until clear; then pour off the clear liquid for use. Do not use any part of the sediment. For ordinary use 10 grains of bicarbonate of soda (common baking soda) to each pint of water, will make a better alkaline water than the lime. This should be used in constipation. If neither of these methods should prove satisfactory, refined gelatine, such as the Keystone (made by the Michigan Carbon Co.) may be soaked in twice its bulk of cold water until soft, and then boiled and strained. As gelatine is a good and easily digested food, a considerable quantity may be added to the water, to be used for diluting the milk."

"Which of these diluents is preferable?"

"The barley water or gelatine, the bicarbonate of soda for sour stomach, and the lime water in case of diarrhoea; but neither lime water nor soda should be used continuously."

"These diluents are for breaking up the curds. How do you make the milk richer?"

"That is done by adding cream and sugar. Sometimes half cream and half milk are used, but it is better to take the top milk; that is, after the milk has stood some six hours the cream and milk of the upper half of the can or jar is skimmed off for use. To each half-pint of top milk two and a half to three heaping teaspoonfuls of ordinary sugar should be added."

"How would you mix the ingredients for a child two months old?"

"A child two months old would require, say, twenty-four ounces each day, prepared as follows:

Top milk .... 8 ounces.
Barley water .... 16 ounces.
Sugar .... 4 heaping teaspoonfuls.

If gelatine water, lime water or soda water be used instead of barley water, it will require the same amount."

"Will children thrive better on this mixture than pure milk?"

"Very much. Milk, without dilution, is too rich for many babes. They cannot digest it, and are not nourished, but get diarrhoea and die. Many a young babe has been carried to the grave, because its mother did not know of this way of modifying milk."

"What is the next most important thing to know?"

"That the milk has not been poisoned by disease-breeding germs."

"I don't see how we are to know this."

"If people do not keep their own cows, so that they know that they are clean and healthy, and do not know that the water used in washing the milk vessels is not contaminated by barnyards or privies, or if so, that it has been boiled before using on milk vessels, it is not safe to use milk unless pasteurized or sterilized."

"How about cellars with decaying vegetables?"

"Well, milk must be kept in an atmosphere that is sweet; if this cannot be done, it must be put in sealed or air-tight jars."

"I have often heard that milk is a great absorbent of poisons from the atmosphere."

"That is true; for there is no other animal food which so quickly decays as milk, or which so readily absorbs poison from the atmosphere, so that the greatest care is needed to prevent its contamination. Milk is an ideal food for infants and children, but if not kept from infection, it becomes a source of virulent sickness and death. This fact makes it incumbent upon us to use the utmost care to protect milk from all unclean or contaminating influences, and it must never be allowed to stand in open vessels, where there is foul air, and especially in the sick chamber. It is even objectionable to have milk stand in open vessels in sitting rooms, kitchens or pantries."

"Suppose there is doubt about the quality of the milk?"

"There may be doubt if you keep your own cow, and there certainly will be if milk is purchased from dairymen. In such cases, it will be much better to buy pasteurized milk in bottles, which should be kept tightly corked. If this cannot be done, the milk should be strained and pasteurized in bottles or fruit jars that are fitted with airtight lids; the latter are preferable, because easier cleaned. (For method, see 'Milk.')"

"Doctor, I have heard that it is better to feed a child with milk from only one cow."

"Yes, many writers have advocated this; but it is more reasonable to suppose that the milk from a herd of cows would have a more uniform daily average than that of any one cow."

"Should the milk be warmed before giving to the child?"

"Certainly. Enough should be poured out of the supply jar for one feeding and the bottle set in warm water (not hot enough to scald the hand), and left until the milk is as warm as fresh milk.'

"How much should be given at each feeding?"

"That depends on the age of the child: you will see from the table I gave you that at first a baby takes only an ounce of milk at a feeding, but when a year old, eight or nine ounces at a time. One thing is of greatest importance, and that is, not to put more milk into the nursing bottle than the child should have at one feeding, according to age and amount given in the table."

"Why is this?"

"It prevents over-feeding, and you know exactly what the child is getting. If there be indigestion, the amount should be at least temporarily reduced; and if extra hearty, slightly increased. There must be uniformity, both in amount and as to time."

"This can't be done when the mother nurses her babe?"

"Yes, it can; the child should not be permitted to nurse longer than fifteen or twenty minutes. Some foolish mothers are disposed to give their babies everything they want, as though their opinions were worth more than the most learned men, who have cared for thousands of children, both in hospitals and private practice. The safe side is on that of short allowance; this will not likely do any harm—extra allowance probably will."

"If one lived in a city and found it difficult to get any milk except what is partly skimmed, what should be done?"

"Some sweet cream should be purchased and mixed with the milk—say one part cream to two parts of milk. This should then be diluted with barley or gelatine water, freshly made, put in a bottle or fruit jar, then pasteurized and set in a cool place. The amount necessary for each day should be prepared in this way."

"What is the best way to feed babies with milk?"

"The nursing bottle is generally used, and it is one of the most objectionable things connected with hand-feeding."

"For what reason?"

"From the fact that bottles are hard to clean, and because people persist in using rubber tubing. This can hardly be cleaned, and is, therefore, a breeding-house for bacteria."

"Then it is better not to use any tubing at all?"

"Well, no mother can afford to have disease-breeding tubing attached to the nursing bottle if she wants her babe to live. The bottle should hold about a half-pint, should have a sloping neck and oval bottom, that it may be easily cleaned with a brush or sterilized cotton. The nipple should be attached direct to the neck of the bottle and be so constructed that it can be turned inside out and thoroughly cleaned. The bottles should be washed in borax water and then boiled."

"Some people will say that all these precautions are a good deal of trouble."

"That is true; but not half as much as a sick baby. Those who would rather have their babies in the cemetery need not take the trouble."

"Will the method of feeding you have outlined insure healthy children?"

"As a general rule it will, but not always, and when milk disagrees, other methods are resorted to. It sometimes happens that the prepared foods will agree with a child when milk will not."

"Are the prepared infant foods made of milk?"

"Well, there are milk preparations, such as Horlick's malted milk, but most of the prepared foods are made of starch, dextrinized, or partly digested, by diastase, or other methods. They sometimes serve a good purpose, but even though they make a child fat, they are seldom healthy."

"In the event that milk disagrees, what is to be done?"

"That may happen because the child gets too much or too rich milk. In such cases, a less quantity should be given, or the milk may be reduced by adding a little more

water, and not so much sugar.   If the child is not sick,
but does not thrive, the milk may not be rich enough.
It must be remembered that infants do not all require
foods of equal richness or the same ingredients, and that
milk varies much, depending on the breed and the feed
of the cows. In the Summer, milk is richer in dry weather
than in wet, because the grass is drier and richer."

"Is it advisable to give meat broths or other foods to
young children?"

"Yes; broths made of lean beef, chicken or veal, may
be used instead of milk, for short periods, when there is
indigestion or diarrhoea.   They should be made by macer-
ating chopped lean meat in cold water and then pressed.
The juice should be warmed, but not boiled.   Cold water
absorbs much more of the nutritious part of meat than
hot.   Broths made with hot water are not nourishing.
Some children thrive on cream gruel."

"How is it made?"

"Take rolled oats and add three and a half times its
bulk of cold water.   Boil an hour and a half, or until it
is dissolved to a pulp.   Strain through a fine colander
(sieve) while hot (the strained portion should be about the
consistency of jelly when cold).

"To the strained oatmeal add an equal part of sweet
cream and one or two teaspoonfuls of sugar; then add
three to four times the bulk of both oatmeal and cream
of boiling water.   This should be an admirable food for
children eight or ten months old, although children five
months old have done as well as they could possibly have
done on any food."

"Why is it objectionable for children under eight
months?"

"It is claimed that young babies do not digest starch,
and some eminent authorities say they should not have

starch before they are ten months old; others equally good, say that the ability to digest starch commences to develop when the child commences to grow and increases so that it is permissible to give starchy food at six or eight months of age. For very young infants the cream gruel should have malt, or diastase to digest the starch before feeding."

"About what age should children be weaned?"

"It is always advisable for them to nurse through the second summer, if the mother's health permits it, although it is sometimes necessary to wean children very young. At any rate, the weaning should not be begun during the hot season, if it can be avoided, nor under a year, or over eighteen months old."

"Should the nursing be suddenly stopped?"

"No; they should be fed cow's milk, modified as directed. It would be better to try two parts water to one of top milk in the beginning of the weaning period. As the child grows, the water may be reduced to one part, instead of two. The milk-feeding should take the place of the mother's nursing at same regular intervals, and the nursings should be dropped gradually, and the weaning cover a period of two months."

"Should children ever be bottle-fed and nursed during the same period?"

"Whenever a child does not thrive, bottle-feeding should be tried for some of the feedings instead of nursing."

"When may a child be given foods other than milk or gruel?"

"Strained meat broths may be given at almost any age, and next to it is soft-boiled eggs, or eggs stirred into hot, but not boiling, gelatine water. A child cannot masticate solid food until it has teeth, and milk, with sugar, beef or chicken broths, soft eggs, bread and milk, and cereal or starch gruels must form the essential part of every child's diet, until it has teeth. The practice of giving young children solid foods like meats, raw vegetables, raisins and like substances, has been the instrument of death for thousands."

"Are fruits not permissible?"

"Sour fruit juices are not permissible with milk; but fruits, like apples or peaches, when cooked and free from solid substances, may be given children over a year old."

"How many meals a day should a child receive when it commences to eat such foods as you have named?"

"From four to five meals a day during a child's second year."

"Should children be given tea or coffee?"

"Young children must not be given tea, coffee, beer, liquors, or fermented drinks of any kind."

"What foods may be given children over two years old?"

"I will first speak of some of the foods not to be given them. The worst abused children are those who are indulged by their parents to such an extent as to be allowed to eat everything they see. They must be kept out of the pantry; for nothing could be worse than permitting them cakes, sugar, pastry, green fruits, or anything else they may happen to want. Besides the objection to such articles, they are frequently allowed to eat them at all hours of the day, and if life were not so exceedingly hard to destroy, the mortality rate of children would depopulate the country. Parents are disposed to be particular about almost everything for their children except their diet, and in this they are less restricted than grown people, although they are in greater need of it."

"This is not very definite about what they should eat."

"But I have only discussed preliminaries, and I am going to strike out their ordinary diet at one blow, by throwing out the frying-pan and all fried food; nor will I stop here; pickles for children are instruments of death, but are not worse than sourkraut, and not much worse than griddle or pan-cakes, salads and raw vegetables."

"Which of the fried foods are the worst?"

"Fried eggs and fried salt pork, ham or shoulder."

"Doctor, your attack on the ordinary way of feeding children is rather sharp, and your list of prohibited foods rather sweeping."

"That may be, but there are still more; pepper, mustard and all condiments, except a small amount of salt, must go; together with cheese, bananas, cherries, grapes (unless skins and seeds are removed), blackberries and raspberries, except the juice, gooseberries, cranberries, currants, stringy vegetables, unless chopped fine, canned fish, hot, doughy biscuits or bread, cakes, pies, doughnuts, nuts, unless ground, popcorn, raisins for children under eight years old, the skin of fowls, green or over-ripe fruits, tomatoes, muffins, fritters, salt fish, peas and beans, unless ground or thoroughly cooked and passed through a sieve, green, dried, or canned corn, new potatoes, ice water and ice cream, except in small quantity, when slow eating can be enforced."

"Doctor, your lists are as sweeping as a cyclone. Are there any foods except milk you haven't condemned?"

"Plenty of them. There is bread, cracked wheat and wheat foods, corn preparations other than green, dried, and canned corn, rice, oatmeal, barley, rye, meats in small quantities, boiled or roasted, eggs, raw or slightly cooked, fruits, except as prohibited in the list given, fresh fish, cooked vegetables, when strings are cut very short, baked or mashed potatoes, arrow root, tapioca, sago, and gela-

tine.   Wheat, oat and corn mushes should be strained for
children under five years old."

"What ought to be the staple diet for children?"

"Milk, entire wheat bread, oatmeal, wheat gluten, grits,
or germ meal, fruits, rice, meat, fresh fish, and soft-cooked
eggs. In all food preparations or mixtures, it must be
kept constantly in mind that all the starchy foods require
much cooking, while meats and eggs but little; also an
excess of fat and sweets must be avoided. As an example, if eggs are used in rice pudding, they must be added
after the rice is cooked, for there will be enough heat in
the rice to cook the eggs."

"Should the diet of growing children differ materially
from that of older persons?"

"It should contain more tissue-forming food and more
mineral matter. These elements are found principally in
milk, wheat, oats, meats and eggs. The first three in
some form or other should compose the main part of
their diet. Growing children who do not have foods containing lime, will do better on hard water than on soft, as
the former furnishes lime necessary for the bones."

"Doctor, can you give a model diet for different ages
of children?"

"It is easy to indicate suitable food, but very difficult
to be definite as to quantities, because there are so many
modifying circumstances. One child may be as large and
active at three years of age as another at five; then temperature, clothing, exercise and growth are all elements
of food requirements. Probably as near an estimate as to
quantity of food needed for an adult is about one ounce of
food as ordinarily eaten, or half an ounce of dry food, to
each three pounds of weight of the individual, but for fat
people, or the sedentary, this would be much too high
for a daily average. Now, as I have already stated, chil-

dren require a higher per cent. of nitrogen, because, in addition to the ordinary waste of tissue, they must have something for growth; but as the average growth does not exceed one-third of an ounce per day, which is seventy per cent. water, it will be seen that the need for growth has been greatly exaggerated by many writers. A child a year old will consume forty ounces of milk, containing five ounces of solid food, while the average growth of a child per day will not exceed a tenth of an ounce of solid matter."

"How would that compare with the standard diet of grown people?"

"The diet for grown people, weighing seven or eight times as much as a child one year old, would contain about four times as much protein as the diet of an infant one year old."

"Doctor, how would you divide the different periods of a child's life, for the purpose of arranging dietaries?"

"The first period is from birth up to eight or ten months of age. During this period modified milk is next best to that of the mother's. When these fail, pre-digested starch preparations, sold as prepared foods, such as Imperial Granum, should be tried. Some children do better on them than milk, and some give them with milk to great advantage. Part of a beaten egg may be given for temporary use."

"Then when a child is eight or ten months old, it may be fed some starch?"

"Yes; white bread, crackers, arrow root and sago may be added to the milk given the child. If bread be used, it must be good bread, well baked and dissolved in milk or hot water, and given in small quantities. Wheat, oats, and rice preparations, when boiled to a pulp and strained through a very fine sieve, are very useful additions to milk."

"When may other foods be given?"

"It must be kept in mind that a child must not have solid food until it has teeth, although other soft food, such as mashed potatoes, baked or stewed, sweet or sub-acid apples, free from peel, seeds and core, may be given."

"May other fruits be used?"

"They must be used with great care; all very sour and astringent fruits must be avoided. In constipation, slightly acid fruit juices, when strained, may be given two or three hours after meals, and one hour before."

"When may other animal foods be added?"

"Soft-boiled or poached eggs may be given children in their second year, and in some cases the first. It would be well to give only one or two teaspoonfuls at first, and never more than one egg at a meal until a child is four years old. The practice of permitting children to eat two or three hard-fried eggs is most reprehensible and dangerous to the child."

"When may solid food be allowed?"

"A child should have a good number of teeth at two and a half years of age, and this may be said to be about the beginning of the third period."

"Are no meats to be allowed before a child is two and a half years old?"

"Meats are given after eighteen months of age, but they must be scraped, ground or in some way reduced to a pulp or powder."

"After a child has teeth, I suppose it may be given a dry diet?"

"Only to a limited extent. The diet should still continue much the same, except that the bread need not be soaked, nor the meat powdered. Cooked garden vegetables (one variety at a time), chopped crosswise of the fibre, may be added to some of the meals."

"When would you change this diet?"

"Well, there should be no radical change made from this diet, except an increase in quantity, and some relaxation as to straining foods, when a child reaches five or six years of age."

"When would you allow such prohibited foods as tomatoes and bananas?"

"They might be tried in a limited way, at six or seven years of age; baked bananas at two or three."

"Would you allow the use of fried foods at this age?"

"No; I would bar the frying-pan for all ages."

"Doctor, you seem to be severe. You must consider the effects of bad feeding and training very far-reaching in their effect."

"The fearful infant mortality only faintly indicates the direful results of ignorance on this subject. Who can measure the sorrow, anxiety and care expended on sick children, that could easily be avoided? Nor is this all; they are allowed to grow worse than maimed, a burden. to themselves and often a care on their friends or society. Why is there not some anxiety on the part of parents, to give their children freedom from pain and disease, as well as riches? Is not a sound body more conducive to happiness than wealth?"

"Then you think if children were properly ushered into manhood and womanhood, and taught how to live, most of our troubles would be averted?"

"Undoubtedly; even a weak child, if properly fed and trained, may be developed into good, healthy manhood or womanhood, and their growing period is the time to correct their defects."

From 12 to 18 Months Old.

"Doctor, will you arrange dietaries for children from the age of one year to maturity?"

"I have already done so, and will read it to you:"

"A child 12 months old should be fed at about 7 and 10:30 a. m.; 2:30, 6 and 10 p. m. If the child is not weaned, it will probably be advisable to allow it to nurse the first, third and last meal, and fed the second and fourth. When the nursing is reduced to twice a day, it will be best to nurse the second and last meals, and finally feeding may be substituted for these, as weaning progresses. A child a year old, will require forty ounces of modified milk, one-third of which is milk and cream—'top milk.' A child a year and a half old will require a pint, to a pint and a quarter of top milk, and two or three ounces, when strained, of well-cooked starch, either rice, barley, flour, arrowroot, sago or oatmeal, four or five teaspoonfuls of sugar, and a pint and a half of water, for the five daily feedings. Meat broths, egg or prepared foods, may be substituted if they agree with the child better than milk and starch."

"In following this outline for feeding, what would be the most probable error?"

"Giving an excessive quantity of food and too little fat —the result of poor milk."

Dietaries—1½ to 2½ Years of Age.

Milk, cereal gruels and mushes, sago, arrowroot, tapioca, eggs, bread and milk or broths, scraped meat in small quantities, meat broths, rice, milk or gelatine or starch pudding, stewed fruits that do not require sugar, such as apples and prunes, without skins.

2½ to 6 Years.

To above add: Meat, powdered or scraped, bread, entire wheat, fish, fruits, according to directions, cooked garden vegetables—except tomatoes, cucumbers and peppers—wheat gluten, mashed potatoes, baked potatoes.

6 to 10 or 11 Years.

Additional foods: Tomatoes, bananas (occasionally), raisins, oysters. Straining will not be necessary for cereals, but for legumes, peas, beans and lentils, ground or cooked until they are of consistency of puree, powdered nuts."

CHAPTER XXI.

DIET IN PUBERTY.

"Doctor, why do you make a division at 10 or 11 years of age?"

"The dietary from 6 to 11 was intended to reach to the age of puberty."

"Then you regard puberty as a critical period?"

"For girls it is extremely so, because mistakes at this time not only seriously affect girlhood and womanhood, but it also curses unborn generations."

"How is that?"

"The young girl of to-day will soon be the mothers of another generation, and what affects their health will likely affect their progeny."

"What connection has food with such dreadful results?"

"There can be no growth without suitable food; for nourishment is a vital element of all life. Now, when a girl reaches puberty, there is an increased physical demand, for two reasons: (1) It is a period of more rapid growth, or at least it should be so. (2) The functional development of the sexual organs causes an increased drain on the system, which, if not met by suitable nourishment, results in injury well-nigh immeasurable."

"Is that the reason why young girls are so often anaemic?"

"It is the principal reason. A girl cannot grow into healthy womanhood without good blood, and if she has it not, the effect is as obvious as a long drouth on the summer harvest."

"Do you mean to say, that the disorders peculiar to

women, with the agony they have to endure, are mainly due to lack of care during puberty?"

"They are largely due to lack of intelligent care between the ages of 11 and 17. Many girls receive a kind of well-meant care, that is worse than total neglect. They are the children who are fed dainties, over-dressed, restrained, and in winter kept in rooms ten or fifteen degrees too hot; but in summer are dressed in the thinest fabrics, no matter how cool the weather. Woman's physical woes can be described in short terms: Idiotic feeding, and maniacal folly in dress."

"That is strong language."

"But not too strong. An idiot is a person without reason. When we do things without reason, things, too, that dumb animals will not do, are they not idiotic? Now, as the conventional dress of women is responsible for a large per cent. of their ills, what less can we call it than mania?"

"But how is dress related to feeding?"

"In this way: A well-nourished body, to a great extent, protects itself; but if the organs of the body are displaced, or the circulation interfered with by tight clothing, it cannot do so."

"Be a little more specific, doctor. Name the habits that seem to you the most injurious."

"Eating at all hours of the day. Eating improper food, such as pop-corn, cake, candy, pickles, green and over-ripe fruit, fried foods and doughy bread, saturated with butter or gravy. During puberty, girls' appetites seem to crave all sorts of things, because they see others eat them; whereas, the demands of the body require food rich in tissue-forming substances, and not very difficult to digest. Eating between meals is one of the most pernicious habits of school girls, and it can't be cut too short. Pampering children with all sorts of pastry and highly-seasoned

dishes, destroys the taste for natural food, and curses them
for life. They should be fed on plainly, but well cooked
cereals, well-baked bread, from entire grain, milk, meat,
eggs, cooked without fat, and sound, ripe fruits. A limited amount of sugar, syrup or candy, may occasionally
be eaten at meal time. Pop-corn and nuts are wholesome
when finely ground, but must be prohibited as ordinarily
eaten."

"Young people should have good digestion, why so
particular?"

"Because the newly-developed functions of sex interfere
with digestion for about five days before and after stated
periods, so that nearly a half a month is taken up with the
excretion of waste and repair, which makes them extremely sensitive to cold and liable to constipation, both
of which must be shunned as deadly enemies."

"Why should they shun constipation more than other
people?"

"Well, besides the importance of good digestion at this
period, accumulation of fecal matter in the bowels, disturbs the circulation in the delicate organs of generation,
and may cause a life of suffering."

"Doctor, you seem to favor both freedom and restraint."

"Yes, a girl should be dressed so as to allow the greatest personal activity, and mothers should remember that
a daughter's health is far more important than lady-like
deportment. As an example of anaemic women, there
are none so bad as the French of the upper classes. Restraint, convent life, and folly in dress, make the French
women the poorest, physically, that exist in any enlightened country."

"How would you overcome the disregard for warmth,
nourishing food, regular eating, and lack of exercise?"

"By teaching girls before they rich puberty, that they

are to become women, and that it would be far less injurious for them to cut off an arm or a foot, and less painful, too, than to be badly developed women and have to suffer all their lives."

"How about diet for boys?"

"If fried foods, green and over-ripe fruits, and an excess of food, be kept from boys, they will not be sick."

"How can over-eating be prevented?"

"By taking all the food necessary for one meal on the plate or dishes at one time. Boys should not be allowed to repeatedly help themselves, for no attention is paid to the great quantities of food eaten in this way."

"Doctor, you have indicated from your remarks that you were a strong believer in pre-natal influences, and I suppose that diet and the mental conditions of prospective mothers are very important factors in shaping the character of unborn children?"

"Undoubtedly; the unborn child is mainly dependent upon its mother for its physical life, and to a great extent its mentality, and these, in turn, must have proper nourishment or be undeveloped."

"Are there not other influences which affect the prenatal life of the child?"

"Yes; this is especially true of dress. A well-known author, when asked when a prospective mother should discard the corset, very pithily answered, 'Two hundred years before her child is to be born;' but this does not belong to foods."

"No, I am sorry; but in the companion volume you can sing the undying dirge of the waist-constrictor and pain-producer of female apparel. But what are the faults of the mother's diet that make her children so imperfect?"

"The unborn child receives its nourishment direct from its mother's blood, necessitating good health on her part."

"Is there any particular kind of food required?"

"There is a theory advocated in Tokology, and other books, that child-birth is made easy by a fruit and starch diet. It is argued that acids dissolve mineral matter and prevent the bones of the unborn child becoming solid, and that when fruits are used in connection with foods containing but little lime and other mineral substances, the bones of the child at birth will be extremely flexible, and birth, therefore, very easy."

"You don't endorse the theory?"

"No; because the bones of all children at birth are soft, and when they are deficient in lime, the child will be in a diseased condition called rickets. In health, nature always preserves its own balance, and when it cannot do this we have disease."

"Probably the good effect is due to the fruit diet?"

"That is it. A wrong theory did not spoil the good effects of the diet when it happened to be particularly adapted to the person using it."

"Then you endorse the fruit and starch diet, but not the reason given for it?"

"Not entirely—it is a good thing carried too far. As already explained, fruit is an internal cleanser, which gives life and elasticity to the tissues of the body and prevents constipation and uric acid concretions, and it is these effects which have given such satisfactory results to prospective mothers."

"What is it that you condemn?"

"It has several faults, chief of which is the indiscriminate use of fruit acids and starches. To give you an example of the effect of acids and starch, I recently emptied a man's stomach eighteen hours after eating tapioca. Now, tapioca is practically pure starch and easily digested, but in this particular case there was an excessive secretion of acid, and the tapioca was not digested in eighteen hours, but the particles were much larger than when swallowed. In another case, I found undigested and unchanged grains of rice five hours after the lady had eaten rice and two oranges."

"Might that not have occurred with meat?"

"No; I have emptied stomachs where there was excessive acid secretion and found meat digested within one hour from the time when eaten."

"Then you favor a meat diet?"

"Only to a limited extent. I favor a fruit diet, but not such incompatible foods at the same meal as rice and oranges and rhubarb and toast."

"You must know of some ill effects to unborn children?"

"Yes; excessive or imprudent use of fruits derange digestion and bring dyspeptic and crying babies into the world."

"Some people think they cannot have too much of a good thing."

"I am not one of them. It has been my constant study to find out the use of foods from a practical standpoint, rather than follow the speculative theories of either scientists or 'faddists,' and my original investigation makes me an enthusiast in the use of fruit."

"What diet would you advise for the pre-natal development of the child and the health and comfort of the mother?"

"The welfare of the unborn child and its mother are inseparable. Her largest meal should be breakfast and her lightest one supper. The daily diet should consist largely of fruits and cereals—wheat, oats, and rice, with entire wheat bread for the staple part of the diet. Broiled or stewed chicken, baked fish, broiled, boiled or roast beef or lamb may be eaten for one meal, breakfast or dinner, on alternate days. The meat must be powdered by grinding or great care taken in its mastication. Fried or tough meat must be wholly excluded. One or two soft-cooked eggs for breakfast or dinner may be eaten on alternate days, when meat is not allowed. The general rules laid down for the use of fruit apply to all conditions. A model dietary would be something like the following:"

Pre-natal Dietaries.

Breakfast—For tissue-forming foods use one or two of the following, according to taste and convenience: Eggs soft-cooked at low temperature.

Fresh beef, mutton, chicken, venison, quail, pheasants, stewed or roasted—no canned or salt meats—fresh fish, boiled or baked, oysters (fish and shell-fish are so often contaminated that they are more or less dangerous), peas and beans ground and thoroughly cooked, or boiled and passed through a colander, powdered nuts or nut foods, wheat gluten, milk when not used with sour fruits, so as to form large curds.

For Starches.

Dry toast, dried and then browned by hot coals or very hot oven. Roast grains that have been well boiled before roasting.

Vegetables to Suit.

Stewed celery, boiled onions, stewed asparagus, spinach, well cooked, tomatoes (occasionally), squash, lettuce, string beans, green peas, radishes (only in small amount when in good health), rhubarb (occasionally, in small amount).

Fruits.

Sour fruits should be used with the meal containing the least starch; for that reason we class them with the meat or egg meal. Baked or stewed apples, such varieties as

Ben Davis, Wine Sap, Northern Spy, and Bellflower. Oranges may be eaten at breakfast or an hour before, with small cup of hot water. Grapes, without skins and seeds, strawberries, plums of the large varieties, but not the astringent kinds, peaches, pineapple juice, but no fibre.

Fats.

Cream, butter, nut butter, powdered nuts or nut foods, breakfast bacon—broiled. Butter is often more or less rancid, and is worse in this respect than cream. Good cream and nut butter are the best of all fats.

Drinks.

Not more than four ounces of fluid is allowable of one of the following:

Hot water, hot water and milk mixed, caramel cereal, cocoa and chocolate in small quantities are permissible where there is active exercise. Breakfast should contain from one and a half to two and a half ounces of protein—tissue-forming food—and should give from one thousand to fourteen hundred calories. (See tables giving composition of foods.)

Dinner.

Dinner should not, under any circumstances, be less than five nor more than seven hours after breakfast, and should be regular. Six hours is the best, and may include one or two articles from the following list, for each meal:

Corn bread, whole wheat bread, Ralston Health Club Breakfast Food, wheat germ grits, Granola, Crystal Wheat, rolled oats, rice, beans, hominy and other cereal foods. All may be served with milk.

Vegetables.

Potatoes—baked, boiled, stewed, roasted or mashed—though mashed potatoes are objectionable, because they do not get sufficient saliva in eating, and become too easily swallowed. Boiled cabbage, without fat, celery, raw or stewed, greens, spinach, cauliflower, pumpkins, squash, green peas, string beans, green corn, tomatoes.

Fruits.

Apples—sweet or sub-acid, baked or stewed and eaten without sugar; peaches that are not rich in acid, sweet grapes, figs, stewed, dates, stewed with skins removed, pears (with exceptions of those that are puckery—they are astringent and not allowable), prunes with skins removed.

Fats.

Same as breakfast.

Dinner should furnish one or two ounces of protein and about twelve hundred calories of heat. This is not arbitrary, but a guide to diet properly balanced.

Supper.

Stale, dry bread, dry or milk toast, boiled rice—preferably boiled and roasted—wheat foods, tapioca or sago, baked potatoes, honey and molasses (sparingly), baked or stewed apples, sweet grapes, watermelons.

Dessert.

Fruit pudding, custard, corn-starch, rice pudding, gelatine pudding, ice cream, in small quantities, slowly eaten.

Fats.

Same as breakfast, only in less quantity.

Drinks.

Milk, if it agrees, otherwise same as breakfast. The breakfast meal may sometimes be made the dinner (noon) meal, and the dinner meal the breakfast. Sugar should be avoided so as to allow the largest use of fruits and starches. The astringent fruits, such as blackberries, raspberries, dew-berries, cranberries, pomegranates, wild cherries and quinces are to be avoided, except when there is a tendency to diarrhoea. If bowels are too free, leave off the coarse vegetables, the cereals containing bran, and the sour fruits. The general rules heretofore explained should govern. The amount of food must be adapted to the needs as governed by size, exercise or labor, weather and peculiarities. Prevent constipation without drugs.

CAUTION.

Never eat many different foods at one meal. Three different foods at one meal are better than a large number. Craving very unusual or unseasonable foods is unnatural. Keep the thought of food, and for that matter all thoughts of self out of mind. It is of greatest importance that the will be exercised to keep well and pleasant and not be disturbed by the disagreeable things of life. The mind should be occupied, in a useful way. If there be great desire for something unusual it should be gratified in such a moderate way as not to do harm."

DIET IN CONFINEMENT AND FOR NURSING MOTHERS.

"The bringing of a new life into the world is a great responsibility, and as the health and character of the child is dependent upon its parents, the time must be near when they will see that it is far more important to have children that are fit to live, than it is to leave them wealth. In ordinary cases, no food will be needed during labor, but in protracted cases it is better to sustain the strength by a cup of hot meat broth or hot milk. It was formerly thought that puerperal women should be fed for several days on broths and gruels, under the belief that it kept down puerperal fever, which was much more common, before the danger of bacteria was known, than it is now under modern aseptic surgery. After her delivery, the mother should drink water freely, and after a few hours' rest she may then be given a cup of hot bouillon or other meat broths, but they must not contain a large amount of fat. If the patient is disposed to eat anything, she may again be fed a small amount of milk toast, in four or five hours, if made according to directions. Some physicians allow meat and solid foods, but it would seem to be better to confine the diet to soft and easily digested foods until the bowels have moved two or three times. Among the foods allowable for the first two or three days, are: broths, milk if it agrees with the patient ordinarily; one egg at a meal, if cooked but little, without fat, or an egg may be stirred in any of the broths, only moderately hot, but not boiling. Wheat, breakfast foods thoroughly cooked, boiled rice, cooked four hours, baked sweet apples, cream, a little butter or nut butter, and any of the drinks allowed before childbirth. On the second or third day, she may resume her ordinary diet, unless there is some particular reason for not doing so. After a child is born, a mother has two lives to feed from one set of digestive organs. Her own health must be considered and also that of her child. And in this connection it will be useful to consider what affects the mother's milk. The medical profession believe that acids ingested by the mother, cause colic in her babe and sometimes griping and purging, and therefore forbid the use of ordinary fruits, but the sweet fruits are not only allowable, but beneficial. Potash salts, eaten by a nursing mother act as a diuretic in the nursing child. Large quantities of potatoes eaten by the mother would likely act as a diuretic in the child, but no experiment of this kind has ever been reported, and no apparent injury has ever been observed. The greatest danger is from the indigestion of the mother. The human system being a sort of laboratory, if it be thrown out of balance, it may poison itself, and some of the poison must necessarily appear in the mother's milk. Violent exercise or great emotion or mental strain on the part of the mother endangers her nursing child. This is not all the danger to which the child is subject, for an overdose of laudanum taken by a mother has been known to kill her nursing child.   Antimony and iodide of potassium are said to pass most readily into milk, while senna, rhubarb, sulphur, castor-oil, turpentine, copaiba and anise; the salts of mercury, lead, arsenic and zinc are excreted in the milk.   Nursing mothers must be careful about taking drugs.   Diet must be adapted to secure good digestion, and constipation must be avoided by proper regulation of diet, which should be done according to the rules heretofore laid down.   Menstruation during nursing is likely to change the mother's milk, and make it necessary to feed the child in some other way.   At weaning, the mother should eat a dry diet, and drink as little as possible, to keep in health."

CHAPTER XXII.

DIETETIC ERRORS AND DIETARIES.

"Doctor in discussing digestion and foods you have frequently spoken of dietetic errors. Would it not be a good idea to enumerate them?"

"Perhaps so, but one scarcely knows what to give the greatest prominence. For convenience, I will begin with one of the most general faults, and enumerate them as follows:

1—Overeating.

2—Eating fried foods.

3—Drinking an excess of fluids during meals.

4—Drinking cold drinks at meals or during digestion.

5—Drinking an excess of liquids, especially beer, or ice water.

6—Excessive use of strong tea or strong coffee.

7—Haste in eating, resulting in imperfect mastication, and the insufficient admixture of saliva with the food.

8—Excessive meat eating, including wild game.

9—Excessive sugar eating.

10—Eating doughy bread, pancakes and pastry.

11—Eating vegetables hastily, without chopping the fibres.

12—Eating tough, raw vegetables.

13—Irritating foods, pungent vegetables, pepper, salt, mustard, and other irritating substances.

14—Taking foods and drinks excessively hot.

15—Pickles and vinegar.

16—The admixture of starches and acids.

17—Incompatible foods such as strong tea, and eggs, acids or vinegar and milk, tea, cheese and acids.

18—Eating fruits with seeds or skins, especially black-berries, raspberries, grapes, currants, gooseberries, raisins and cranberries.

19—Eating green and overripe fruits.

20—Excessive cooking of meat.

21—Insufficient cooking of starches.

22—Excessive consumption of fats.

23—Eating too little food.

24—Eating food containing too little waste such as: milk, eggs, white bread, potatoes, butter, sugar, meat.

25—Eating food containing too coarse waste, such as green, dried or canned corn, and the tough skins of peas and beans.

26—Excessive consumption of starch, such as a diet of white bread and potatoes.

27—Diet deficient in mineral matter.

28—Eating an excessive quantity of sour fruits.

29—Foods containing Ptomaines from decay.

30—Eating too many kinds of food at the same meal.

31—Eating too frequently, and not allowing the stomach time to empty itself.

32—Irregularity of eating.

33—Going too long without eating.

"Most of these have been discussed, and those that have not, will be treated at greater length under causes of indigestion."

"Doctor, who requires the most food?"

"Growing boys sixteen or seventeen years old, who do the hard physical labor of mature men."

"What do you call hard labor?"

"Harvesting, clearing land, chopping cord wood, digging ditches, brick and stone masonry, plastering, handling freight and heavy material in foundries, factories and mills, and other labor requiring great activity and use of strength."

"How are the needs of the different classes estimated?"

"You will remember that foods are divided into two general classes: Tissue-forming and heat-producing. The variation of amount of food needed is mainly of the latter class, and is estimated by units of heat called calories or rather by kilogram degree calories."

"What do you mean by calories?"

"Foods have been tested for their heat or force-producing power, by scientific methods, and the term caloric is the unit measure of heat produced. Now one thousand calories make one kilogram degree calorie, which is ordinarily understood when the term calorie is used."

"Then the heat and force producing value of food is estimated according to the amount of work or exertion it sustains measured in calories."

"Yes that is it. Fats are the greatest heat producers, and butter produces 220 calories to the ounce, and tomatoes only five. Next to fats sugar and starch produce the highest calories—flour producing 103, and sugar 113 calories to the ounce."

"What should the diet contain?"

"Our daily diet should contain three to six ounces of protein, and heat-producing material to make from two thousand five hundred to 5,000 calories (exceptional cases may require more than 5000 or less than 2500), mineral matter, and some waste—a large amount for the sedentary and constipated."

"I see the value of this. One could not eat enough cabbage or tomatoes to produce one-fourth enough calories for a hard day's labor."

"That is true, and here is where the vegetarian diet has failed, when relied on for hard labor, because it made

too much bulk to produce the calories necessary. Butter, ground nuts or meats, must be added to a vegetable
diet to raise the calories without making too great bulk.

"Old age is almost synonymous with physical discomfort and disease, as if it were not enough to see the light
of life fading away, nature is inclined to inflict all the over-
due penalties for the transgressions on her for the entire
life. But the aged are not without hope, for such illus-
trious examples as Gladstone and others clearly show
human possibilities. Those who are too thin to cast a
shadow, can scout the idea that they will "dry up and
blow away," likewise the fat rheumatic and gouty, can disprove that excessive fat is but another name for folly.
The digestive organs are often the first to weaken, and
with poor blood, the system is well-nigh defenseless
against disease. Those who have been large eaters,
usually continue so, notwithstanding the lessened demands of the system. This may overcrowd the blood
vessels, which their weakening walls will not stand, and
apoplexy is the result. The most common fault in the
dietetic habits of the aged, is eating an excess of sugar and
meat. This clogs the system with nitrogenous waste, and
causes rheumatism, which is well nigh universal among
the aged well-to-do. Those who would be free from
disease, must bear in mind the lessened needs of the
system that follow from a less active life, common to old
age, and that it becomes less and less able to dispose of
any excess of food. They must also bear in mind that
besides the weakness, incident to advancing years, digestion is also weakened by general inactivity of the body.

WHAT NOT TO EAT.

Fried meat, nor fried foods of any kind. Fresh bread,
as ordinarily made, hot bread, saturated with butter or
gravy, hot biscuit, cakes of every kind, pies with short-

ened pie crust, pickles, vinegar, sauer kraut, salt meats, sausages, salt fish, dried meats, raw onions, raw vegetables, strong tea and coffee.

Foods that may be sparingly (occasionally) used:

Sugar, molasses, syrup, honey, boiled ham, breakfast bacon, sweet potatoes, cabbage boiled without fat, rhubarb, if no rheumatism or disease of digestive organs, astringent fruits, such as cranberries and raspberries.

SUITABLE DIET.

Stale or dry light bread, wheat foods according to taste and condition, but particularly gluten biscuit, or gluten meal, rolled oats, pearl barley, rice, hominy of all kinds, eggs, milk, cream, butter, fresh fish, fresh beef, mutton and fowl, but not oftener than once a day; puree made of peas, beans or potatoes, stewed celery, string beans, cauliflower, asparagus, cooked onions, beets without vinegar; all fruits, except astringent ones, such as raspberries, blackberries and some varieties of pears; quinces.

If old people want to avoid rheumatism, they must avoid eating much meat. They must also be careful about eating fatty foods and sugar, as such a diet will be too fattening, and throw the diet out of balance. For those not engaged in hard labor, two meals a day is all that is permissible. These should be at eight or nine in the morning, and three or four in the afternoon, but should be regular. Nothing must be eaten between meals. If food be needed at night before bed time, a cup of hot milk or a baked apple may be eaten, and will often cure sleeplessness. Tea and coffee are bad for all ages, but particularly so for the aged.

DIET FOR BICYCLISTS AND ATHLETES.

"Athletes desire the greatest strength and endurance, with activity developed in the highest degree. To this end mucles are developed, fat and water reduced."

"How is this accomplished?"

"By a diet rich in nitrogen and poor in fat and starch, aided by systematic exercise, massage and baths."

"As I understand it, the bulk of an athlete's diet in training, is meat?"

"Yes, and if you will notice contests, you will observe that it frequently happens that some one breaks down."

"In the haste to reduce fat, so little water is given, with a diet so rich in nitrogen, as meat is, the kidneys are overburdened, and there is auto-intoxication."

"Then the meat diet is carried to far?"

"Yes, soft cooked eggs and milk are better than all meat, and dry gluten biscuit, without sugar is still better. If good, fresh gluten biscuit are not easily obtained, bread made of wheat flour, or middlings may be washed in cold water until the starch is dissolved, and the remaining gluten may then be baked or cooked as desired."

"What is the advantage of wheat gluten?"

"It serves about all the purposes of meat, without the danger from uric acid, which meat produces."

"Would you allow bread?"

"Yes. Entire wheat bread, because it contains more gluten, or, what is still better, dry crackers made of entire wheat flour without sugar. These are an aid towards maintaining a dry diet and are better than toast. Coarse vegetables must not be used, as there will not be constipation with the necessary exercise, massage and baths incident to training. An orange or half lemon may be occasionally eaten a half hour before meals."

"Then athletes must not eat vegetables?"

"In very limited quantities, if at all. If there be a tendency to constipation, there should be an increased allowance of such foods as granose or cereals with fine bran. Amateurs who have no such aid as massage and baths need more coarse food, and should eat any of the cereals prepared by boiling and roasting."

"You haven't given a complete diet list?"

Stale bread—small quantity; dry toast; beaten wheat **crackers**; biscuit without sugar or shortening; granose, **dry**; bromose; beef steak without fat or butter, or roast beef when cooked by basting in dough; eggs soft, without fat; must not be fried; fresh fish; beans and peas; nut meal; cream; butter; nut butter.

HEAT OR FORCE PRODUCING FOOD.

QUANTITY REQUIRED FOR ONE DAY.

	Light Work.	Moderate Work.	Hard Labor.
Wheat Flour	28 oz.	36 oz.	45 oz.
White bread	38 oz.	48 oz.	60 oz.
Corn meal	28 oz.	36 oz.	45 oz.
Oatmeal	24 oz.	30 oz.	38 oz.
Lard	10 oz.	13 oz.	17 oz.
Rice	28 oz.	36 oz.	45 oz.
Rye	28 oz.	36 oz.	45 oz.
Sugar	28 oz.	36 oz.	45 oz.
Barley	28 oz.	36 oz.	45 oz.
Buckwheat	30 oz.	38 oz.	48 oz.
Beans	28 oz.	36 oz.	45 oz.
Peas	28 oz.	36 oz.	45 oz.
Butter	12 oz.	16 oz.	17 oz.
Eggs	56 oz.	76 oz.	96 oz.
Beef	64 oz.	88 oz.	7 pounds.
Potatoes	7 pounds	9 pounds	12 "
Sweet potatoes	4-5 "	6 "	8 "
Cabbage	15 "	20 "	27 "
Cauliflower	14 "	19 "	26 "
Beets	12 "	16 "	21 "
Carrots	15 "	16 "	21 "

Turnips	18 pounds	2 pounds	32 pounds
Tomatoes	25 "	34 "	45 "
Celery	30 "	40 "	52 "
Onions	12 "	16 "	21 "
Radishes	23 "	32 "	42 "
Cucumbers	40 "	55 "	75 "
Asparagus	23 "	32 "	42 "
Milk	8 "	11 "	14 "
Skim milk	12 "	16 "	21 "
Apples	7 "	10 "	13 "

As all the nutriment, shown by chemical analysis, can never be extracted, this table does not accurately indicate the amount of food required.

Eggs and milk contain the least indigestible matter, while in such foods as cucumbers or pickles it is doubtful if more than half or three-fourths of the nutriment as shown by chemical analysis, is really available for the system. The preceding table is intended to point out the deficiencies of foods as heat or force-producers, and the succeeding one the defects of foods as tissue builders.

Table showing the amount of heat per ounce of the principal foods, and number of ounces of each food from which one ounce of protein can be extracted.

NUTRIMENT IN FOODS

FOODS.	Calories per oz.	Quantity of food from which one oz. of protein can be extracted.
MEATS.		
Chuck	47	5 oz.
Ribs, lean	54	5.2 oz.
Ribs, fat	96	5.7 oz.
Round steak	58	5.1 oz.
Canned beef	88	4.1 oz.
Dried beef	60	2.5 oz.
Veal	50	5 oz.

Lamb	95	5.7 oz.
Pork, shoulders	118	7.6 oz.
Ham	121	6.2 oz.
Salt pork, fat	250	12.2 oz.
Pigs' feet	56	6.2 oz.
Chicken	31	4.4 oz.
Turkey	84	4.8 oz.
Fish	28	4.9 oz.
Salmon	58	4.8 oz.
Oysters	15	16 oz.
Eggs, white		8.5 oz.
" yolk		6.6 oz.
" average	45	7 oz.
Milk	20	30 oz.
Milk skimmed	11	30 oz.
Condensed milk	89	12.1 oz.
Cream	57	40 oz.
Cheese (whole)	123	3.9 oz.
Skim milk	82	3.2 oz.
Gelatine	96	1.2 oz.
Lard	264	all fat
Butter	217	all fat
Oleomargarine	220	78 oz.
Entire wheat	104	7 oz.
Common flour	104	9 oz.
Macaroni	102	8.5 oz.
Barley (pearl)	104	11 oz.
Buck wheat flour	99	13 oz.
Corn meal bolted	103	11 oz.
Hominy	103	13 oz.
Pop corn	117	11 oz.
Rolled oats	116	6 oz.
Rice	102	13 oz.

Boiled rice	56	20 oz.
Rye flour	102	14 oz.
White bread, dry	75	11 oz.
Soda crackers	119	10 oz.
Gluten	24	1.2 oz.
Apple pie	78	30 oz.
Tapioca pudding	49	28 oz.
Beans	99	4.5 oz.
Beans, string	12	45 oz.
Asparagus	7	55 oz.
Beets	13	90 oz.
Cabbage	10	48 oz.
Cauliflower	11	60 oz.
Celery	5	71 oz.
Green corn	22	36 oz.
Greens	17	27 oz.
Lettuce	7	75 oz.
Onions	15	60 oz.
Peas	102	4.1 oz.
Green peas	25	22 oz.
Cucumbers	4	125 oz.
Potato, boiled	30	37 oz.
Sour Krout	9	67 oz.
Tomatoes	12	71 oz.
Sugar	116	

FRUITS.

Apples	21	200 oz.
Bananas	30	83 oz.
Grapes	20	100 oz.
Oranges	14	125 oz.
Strawberries	11	100 oz.
Rasins	102	40 oz.
Figs, dried	87	19.5 oz.

Dates, dried	97	45 oz.
Chestnuts	71	15 oz.
Peanuts	160	4 oz.

APPROXIMATE FOOD EQUIVALENTS.

4 oz.
bread
equals:
{
4 oz. boiled rice and 1¼ oz. round steak,
or 1 oz. chicken, 4 oz. potatoes, 1½ oz. butter.
or 3 oz. corn bread, 9 oz. cabbage,
or 6 oz. boiled potatoes. 6 oz. milk,
or 6 oz.   "   " 1 oz. steak and 1 oz. sugar,
or 4½ oz. cooked rolled oats,
or 4 oz. boiled hominy and 4 oz. milk,
or 2 oz. egg, 4 oz. potato, 2 oz. tapioca pudding,
or 4 oz. potato, 4 oz, green corn, 4 oz. lettuce,
or 4 oz. boiled onion and 4 oz. cucumber,
or 1 oz. ham, 3 oz. rice, 2 oz. skimmed milk,
or 10 oz. milk and 2-oz. cream,
or 2 oz. eggs, 10 oz. apples,
or 2 oz. bread and 12 oz. skimmed milk.

4 oz.
beef
equals:
{
22 oz. skimmed milk,
or 2 oz. beans raw or 4 oz. cooked, ⅛ oz. gluten,
or 1½ oz. peanuts and 5 oz. skimmed milk.

4 oz. fat
beef equals }
3 oz. peanuts.

4 oz.
boiled
rice
{
7 oz. boiled potato,
or 4 oz. green peas and 6 oz. apples,
or 8 oz. cabbage, ½ oz. bacon,
or 3½ oz. cooked rolled oats,
or 1½ oz. egg, ⅔ oz. butter,
or 1 oz. steak, ⅔ oz. butter,   [bacon.
or 6 oz. onions, 6 oz. sour krout, ⅛ oz. lard or

4 oz. boiled
rice, 1-2 oz.
sugar and
2 oz. milk:
}
2 oz. egg and 1 oz. butter, or 1 oz. pork
shoulder and 7 oz. potato.

| 2 oz. milk. 4 oz. rolled oats, (8 oz. cooked) | 4 oz. ham, or 3 oz. lean beef, 4 oz. potato, 3-5 oz. butter, or 23 oz. whole milk, or 4 oz. cooked beans, 7 oz. potatoes, or 20 oz. skimmed milk, 1 oz. bacon, or 3 oz egg, 3 oz. bread, ½ oz. butter, or 2¾ peanuts and 1 oz. potatoes. |

| 4 oz. peanuts | 5 oz. round steak and 1½ oz. fat bacon, or 28 oz. whole milk, or 4 oz. chicken, 6 oz. potato, 2 oz. bacon, or 5 oz. fish and 2 oz. fat pork, or 7 oz. eggs, 1 oz. butter and 3 oz. cabbage, or 4 oz. beans, 4 oz. boiled rice. |

| 4 oz. beans. | 4 oz. beef, 3 oz. pototoes, ⅓ oz. butter, or 5 oz. fat beef, or 3 oz. chicken, 3 oz. potatoes, 1 oz. butter, or 6 oz. eggs, 2 oz. cream, or 4 oz. bread, 2 oz. fish and 1 oz. cream. |

| 4 oz. eggs. | 17 oz. skimmed milk, or 2 oz. lean beef and 2 oz. potatoes, or 2 oz. fish and 4 oz. potatoes, or 3-5 oz. gluten and 1 oz. oat meal. |

| 1 oz. gluten, 1 oz. butter | 5 oz. medium fat beef steak. |

| 4 oz. chicken | 1¼ oz. gluten, or 1¼ oz. gelatine, or 4½ oz. fish. |

| 4 oz. potato. | 4 oz. banana, or 1¼ oz. raisins, or 4 oz. apples and 2 oz. skimmed milk, or 1 oz. green peas and 5 oz. grapes. |

Those foods that are rich in protein, but have but little starch or fat have but few equivalents. The principal foods of this class are chicken, fish, gelatine and gluten.

It will be well to remember that no foods have perfect equivalents; that each food contains more or less mineral matter peculiar to itself; also that protein, starch, fat and sugar are not perfect substitutes for each other. The same foods vary in their composition, so that the proportions that would ordinarily be equivalents, are not always the same.

It must not be assumed that knowledge of the comparative value of foods is of no benefit, for the needs of the human system, come within certain limits, and it is of greatest importance that all persons select their foods to meet their particular needs, and we should be well enough informed to do this, without any special effort, just as a person should be able to speak grammatically, without stopping to consider all the rules of syntax.

Appetite is seldom, if ever, a reliable guide, though it may have been so several thousand years ago.

The cheapest food on which one can live, in most portions of the United States, is corn. It does not furnish a perfect food, but one can live on it for months, perhaps many years.

Ten to 16 oz. of corn makes the cheapest meal and the best cheap meal in the world. The cost would be from one half to three-fourths of a cent, and if perfectly cooked it is quite palatable. Ordinarily it is not half cooked, and to prepare it properly, it should be boiled until the grains will scarcely hold together (corn will require from three to five hours), then dried and roasted until quite brown and dry. It may then be ground or eaten whole, but great care should be taken to masticate it thoroughly, although the boiling makes it dissolve very readily as compared with parched or pop corn.

A little butter and salt improves its palatability. Wheat, oats rye, rice, and barley may be treated in the same manner. After parching they may be softened by a few moments' cooking, but it is best to eat them dry. There are no foods so wholesome and nutritious as well boiled and roasted cereals.

ROASTED CEREALS.

If the cereals were treated as described, and then reduced to fine flour, all the phosphates and gluten would be saved without any objectionable bran, as the cooking and roasting makes it possible to reduce the tough bran to a palatable flour. Starch indigestion would almost be unknown, if dry parched flours were used, for the reason that dry foods cannot be swallowed without mastication and saliva. If people could be taught to use their saliva in their food, instead of trying to float cuspidors and cars, fewer people would be dyspeptic.

The simplest and cheapest diet may be made of cereals or cereals and butter, or cereals and cream, cereals and nuts. A meal would require eight to twelve ounces of dry cereals, one ounce of fat and six or eight ounces of milk. The diet of the Americans is mainly bread, meat, potatoes. An average meal would probably contain about

	Oz. Protein.	Calories
4 oz. bread give	.36	300
4 " meat "	.73	240
4 oz. potatoes	.13	120
1 oz. lard		164
1 oz. butter		217
1 oz. sugar		116
2 oz. milk	.10	40
10 oz. coffee		
Total	1.39	1197

People who eat fried meat and gravies are likely to exceed the above allowance of fat, so that the ordinary meal shows an excess of fat and sugar, but too little waste and too much fluid.

The coarse garden vegetable and fruits are not important factors, in force or tissue production, but they are important for other purposes—filling, cleansing. The main part of our diet must consist of cereal foods, legumes, meats, fish, nuts, fats, starch, potato, sugar, milk and cream. A meat diet for three meals, moderate work, should be about as follows:

BREAKFAST.	Protein.	Calories
4 oz. entire wheat bread, stale	.45	300
2 oz. lean meat	.29	120
4 oz. cooked oatmeal	.33	232
4 oz. whole milk	.13	80
2 oz. cream		114
½ oz. butter		109
½ oz. sugar		59
4 oz. cereal coffee		
Total,	1.20	1014

DINNER.	Protein.	Calories.
6 oz. entire wheat bread	.67	450
6 oz. potatoes	.20	180
2 oz. fat meat, beef	.37	180
4 oz. beans (cooked)	.50	240
4 oz. coarse vegetables	.10	60
1 oz. butter		217
4 oz. milk and hot water	.07	28
Total,	1.91	1347

SUPPER.	Protein.	Calories.
4 oz. entire wheat bread	.45	300
6 oz. cooked rice	.47	330

	Protein.	Calories.
6 oz. whole milk	.20	120
2 oz. ham, boiled	.33	242
½ oz. sugar		58
½ oz. butter		109
Total,	1.45	1159

For those who do no physical work, and take but little exercise, the quantities should be reduced from fifteen to twenty-five per cent, while those doing hard labor will require from twenty to thirty per cent. more in heat-producing foods. The cereals should be slightly increased, but the main addition for hard labor must be in fat and sugar.

The dietary for the three meals is not an ideal one, but made to bring ordinary usage into better harmony with physical needs. It would be better to make breakfast a larger meal than supper, but it is not the usual practice, so the diet list given is arranged accordingly.

We would be doing less than our duty if we did not say that, ordinarily, for those who do no labor, meat should not be eaten but once a day, and by many people not at all. We submit the following as the best dietary for light labor, with meat once a day:

BREAKFAST.	Protein.	Calories.
5 oz. Granose	.71	500
3 oz. milk	.13	80
3 oz. powdered nuts,	.75	480
6 oz. baked apples	.03	126
4 oz. hot water or cereal coffee.		
Total,	1.62	1206

DINNER.	Protein.	Calories.
4 oz. entire wheat bread, dry	.45	300

	Protein.	Calories.
3 oz. roast chickon,	.70	93
4 oz. potato	.20	80
4 oz. string beans	.09	48
3 oz. hominy with 2 oz. cream	.25	165
4 oz. rice or tapioca pudding	.18	310
1 oz. butter		217
4 oz. hot drink		
Total,	1.87	1223

SUPPER.	Protein.	Calories.
5 oz. dry toast	.55	450
8 oz. milk	.26	160
4 oz. cauliflower	.05	44
½ oz. butter		
4 oz. peaches		109
½ oz. sugar		58
4 oz. hot drink.		
Total,	.87	821

For those who eat light lunches at or near noon, the morning and evening meals will be larger, lunch taking the place of supper.

It will be observed that the dietary here given, is considerably below what other writers allow for light work, but to those who are not traveling in ruts already made, it may be learned that the difference between light work and hard labor is much greater than usually allowed. The fault is that the allowance for light labor is too high, and that for hard labor too low. A laborer's meals may be patterned after the following:

	Protein.	Calories.
5 oz. dry bread	.55	375
6 oz. cooked rolled oats	.50	360

	Protein.	Calories.
3 oz. potatoes	.10	90
2 oz. lean meat	.40	116
2 oz. bacon	.10	276
6 oz. milk	.20	120
1 oz. butter		217
½ oz. sugar		58
4 oz. cabbage	.10	52
Total,	1.95	1654

One of the common errors for those who do hard work, is to eat too much coarse, watery foods and to drink too much fluid with their meals. This causes the stomach to be unduly distended, and it is frequently unable to properly handle the great bulk. A moment's reflection will convince anyone that the stomach cannot have the same contractile power when its walls are stretched beyond what they should be, so that when there is large demand for force-producing foods, as in extremely hard labor, it is necessary to eat mostly dry food, and to increase the proportion of fat over that of ordinary diet.

Each person's diet should be adapted to his or her particular needs, and as many people thrive better without meat, it would be well for those to pattern their dietaries after the following:

4 to 6 ozs. whole wheat bread, corn bread or dry biscuits.

3 to 5 ozs. powdered nuts.

6 to 8 ozs. milk.

1/2 oz. butter.

6 ozs. baked apples or other sweet fruit.

Milk, eggs, wheat gluten, peas, beans and nuts, must be relied on to furnish tissue food. For fats, nut meal, nut butter, cream and butter are to be preferred to meat fats.

The cost of butter and cream is against their exclusive use for many people, but it is probable that peanuts will be as cheap as any other food. It is mainly a question of grinding or preparing them.

The addition of fruits and green vegetables makes no great difference in the amount of other foods required. In eating green vegetables that contain a large amount of fiber, regard must be had for its effect on the digestion of other food. If not very thoroughly cooked without fat, then chopped fine and well masticated, such food may remain in the stomach for several hours, until decay sets in.

In concluding the subject of dietaries, the authors would have their readers bear in mind that it is not a subject that can be dealt with exactly as a question of arithmetic. Each person must study his or her requirements in connection with the general properties of foods. Overeating can be prevented by taking what food is needed at one meal on the dishes, and then quit when it is eaten. Do not make a fad of diet, for a large per cent. of the common ailments exist only in the mind. Keep dyspepsia, and all thought of it out of mind, and use some common sense to regulate your diet and habits, and all will be well.

CHAPTER XXIII.

FOOD ANALYSES.

The tables of food analysis here given are made up from many analyses, from many sources, but mainly from the Agricultural Department of the United States:

The percentages given are exclusive of waste and refuse.

BEEF.	Water.	Protein.	Fat.	Carbohytrate	Ash.	Calories per oz.
Brisket (med. fat)	47.4	14.6	37.2		.08	115
Chuck (lean)	71.2	19.9	7.8		1.1	47
(med. fat)	67.8	19	12.3		.09	44
Chuck Ribs (lean)	66.2	18	14.8		1	60
(med. fat)	57.3	17.4	24.4		.09	85
Flank (lean)	66.3	17.7	13		.01	55
(med. fat)	59.8	17.9	21.5		.08	77
Loin (lean)	67	19	12.7		1	56
(med. fat)	60.5	18.3	20.2		1	74
Sirloin (lean)	68.5	19.8	10.7		1	51
(med. fat)	62.1	19.7	17.2		1	68
Neck (lean)	50.4	14.2	5.7		.07	31
(med. fat)	45.9	13.9	11.9		.07	47
Ribs (lean)	67.9	19.1	12		1	54
(med. fat)	55.4	16.9	26.8		.09	96
Round (lean)	70.3	20.9	7.7		1.1	45
(med. fat)	65.8	19.7	13.5		1	58
Shank (lean)	71.5	21.4	6.1		1	41
(med. fat)	67.9	19.6	11.6		.09	53
Heart	62.6	16	20.4	13	1	72
Kidney	76.7	16.9	4.8	4	1.2	33
Liver	66.9	23.1	5.7	3.5	1.5	42
Marrow	3.3	2.6	92.8		1.3	248

Sweetbreads	70.9	15.4	12.1	1.6	50
Tallow	15	4.8	79.9	.03	216
CANNED BEEF.					
Boiled	51.8	24.4	22.5	1.1	88
Corned, cooked	51.2	25.9	18.9	4	80
Dried	45.3	40.1	6.1	12.6	60
Tongue	51.3	21.5	23.2	4	86

It will be well to bear in mind, the fuel value of meat depends mainly on the amount of fat it contains, but that the lean meat contains some fat not ordinarily visible, and that fat meat contains in addition to its visible layers of fat, a great deal more invisible fat than lean meat.

The per cent. of waste in the various cuts of meat is not given, because no one has made any record of how much can ordinarily be gotten out of what may be termed waste. The fore shank is forty per cent. bone and the hind shank fifty-five, while rib has about twenty-five per cent bone, loin thirteen per cent., and round steak about six per cent. bone. Fat meat is likely to be more tender than lean, but less economical, because beef fat is less desirable than many other fats.

VEAL.	Water.	Protein.	Fat.	Car-bohydrate	Ash.	Calories per oz.
Breast (lean)	70.3	20.7	8		1	45
(med. fat)	66.4	18.8	13.8		1	58
Flank (med. fat)	68.9	19.7	10.4		1	50
Leg (average	72.4	20.6	5.9		1.1	40
Loin (average)	69.2	19.5	10.2		1.1	50
Rump	62.6	20.1	16.2		1.1	66
LAMB						
Breast	56.2	19.2	23.6		1	85
Loin	53.1	17.6	28.3		1	95
Shoulder	51.8	17.5	29.7		1	99

MUTTON					
Chuck (med. fat)	50.9	14.6	33.6	.9	106
Flank (med. fat)	45.8	14.8	38.7	.7	119
Leg (med. fat)	62.8	18.2	18	1	69
Loin (med. fat)	50.1	15.9	33.2	.8	106
PORK					
Chuck ribs and shoulder	51.1	16.9	31.2	.9	102
Head	45.3	12.7	41.3	.7	124
Loin (average)	50.5	16.1	32.5	.9	104
Shoulder	47.4	13.2	38.7	.7	118
Tenderloin	65.1	19.5	14.4	1	61
Ham, fresh	62.8	18.5	17.7	1	74
Ham, smoked (av.)	40.3	16.5	38.8	4.7	121
Shoulder, fresh	54.3	15.5	29.4	.8	94
(California Ham.)					
" Shoulder	45	15.8	35.2	6.7	104
Dry salt backs	17.3	7.2	72.7	2.8	200
" " belly	17.7	6.7	72.2	3.4	196
Salt pork clear fat	12.2	4.5	78.8	4.5	250
Tongue	58.6	18	19.8	3.6	73
Feet	68.2	16.1	14.8	.9	58
Bacon, lean	32.7	16.4	45.2	5.7	136
Bacon, fat	18.2	10	67.2	4.8	189
SAUSAGE.					
Pork sausage	38.7	12.8	46.6	1.8	136
Bologna "	59.5	18.6	18.2	2.6	70
Frankfort	55.5	21.7	18.8	3.6	71
POULTRY.					
Chicken	74.2	22.8	1.8	1.2	31
Goose	42.3	13	43.9	.8	131
Turkey	55.5	20.6	22.9	1	84
FISH.					
Black bass	76.7	20.4	1.7	1.2	28
Buffalo	78.6	17.9	2.3	1.2	27

Perch	75.7	19.1	4	1.2	23
Wall eyed pike	79.7	18.4	.5	1.4	23
Pickerel	79.8	18.6	.5	1.1	27
Red snapper	78.5	19.2	1	1.3	25
Salmon	65.2	20.6	12.8	1.4	58
Shad	70.6	18.6	9.5	1.3	43
Sheepshead	75.6	19.5	3.7	1.2	32
Trout	77.8	18.9	2.1	1.2	27
Whitefish	69.8	22.1	6.5	1.6	43
Cod	82.6	15.8	.4	1.2	19
Eels, salt water	71.6	18.3	9.1	1	45
Flounder	84.2	13.9	.6	1.3	18
Herring	72.5	18.9	7.1	1.5	41
Mackerel	73.4	18.2	7.1	1.3	40
SHELL FISH.					
Clams	85.8	8.6	1	2.6	15
Crabs, hard	77.1	16.6	2	3.1	26
Lobster	79.2	16.4	1.8	2.2	22
Oyster, as sold	88.3	6.1	1.4	1	15
Terrapin	74.5	21	3.5	1	34
Turtle	79.8	18.5	.5	.3	25
Shrimps, canned	70.8	25.4	1	2.6	32

The viscera of animals does not greatly vary in composition from that of the animal from which it is taken.

Canned or preserved meats only vary as water and salt are added or extracted.

The canned soups sold in the market contain from eighty-five to ninety-five per cent. water, and from two to five per cent. protein, and are not desirable from any dietetic standpoint.

EGGS.	Water.	Protein.	Fat.	Calories.
White	84.8	12	2	
Yolk	51.5	15	30	
Average	73.5	14.9	10.6	45

The shells of eggs average about ten per cent. of total weight.

MILK.	Water.	Protein	Fat.	Carbo-hydrate.	Ash.	Calories
Milk, Average	87	3.3	4	5	.7	20
Skimmed, average	90.5	3.4	.3	5.1	.7	11
Buttermilk	91	3	.5	4.8	.7	10
Condensed milk	30.5	8.2	7.1	52.3	1.9	89
Cream	74	2.5	18.5	4.5	.5	57
Butter	14.6	1	82.4			217

CHEESE.						
Cheese, whole milk	33.7	26	34.2	2.3	3.8	123
" skim milk	45.7	31.5	16.4	2.2	4.2	82
Pineapple cheese	23	29.9	38.9	2.6	5.6	140
Limburger "	42.1	24	29.4	.4	4.1	105
Gelatine	13.6	84.2	.1		2.1	98
Isinglass, Sturgeon	19	77.4	1.6		2	94
Tallow			100			264
Lard, refined			100			264
Cottolene			100			264
Oleomargarine	9.3	1.3	82.7		6.7	220

CEREALS.	Water.	Protein	Fat.	Carbo-hydrate.	Ash.	Calories
Flour, fine	13.8	7.9	1.4	76.4	.5	102
Entire wheat flour	12.1	14.2	1.9	70.6	1.2	104
Graham	11.8	13.7	2.2	70.1	2	104
Low grade flour	11.4	13.9	2.6	70.8	1.3	105
Spring wheat	11.6	11.8	1.1	75	.5	104
Winter wheat	12.5	10.4	1	75	.5	104
Crushed wheat	10.5	11.9	1.7	74	1.4	105
Macaroni Vermicelli	10.8	11.7	1.7	72.9	3	102
Barley meal	11.9	10.5	2.2	72.8	2.6	102
Pearl barley	10.8	9.3	1	77.6	1.3	104
Buckwheat flour	14.3	6.1	1	77.2	1.4	99
Corn meal, bolted	15	9.2	3.8	70.6	1	103

Hominy	11.9	8.2	.6	78.9	.4	103
Pop corn	10.8	12.3	5.6	71.4	1.4	109
Pop corn, popped	4.3	10.7	.5	70.7	1.3	117
Oat meal	7.2	15.6	7.3	68	1.9	116
Rolled oats	11.2	16.7	7.2	66.8	1.9	116
Rice	12.4	7.8	.4	79	.4	102
Boiled rice	52.7	5	.1	49.1	.3	55
Rye flour	12.7	7.1	.9	78.5	.8	102

Reported analysis of Southern corn shows a very high per cent. of protein—so high in fact, and so different from the authors, that we refrain from publishing any analysis until we have more convincing data.

We regret that we have not more accurate knowledge as to the per cent. of cellulose or indigestible part of the various cereal foods. It appears that fine white flour has less than one per cent. of cellulose; rice about three per cent.; oat meal four per cent.; and corn meal five per cent.; and that the entire grain of wheat, rye and corn contains a still larger per cent. of cellulose.

CEREALS.	Water.	Protein.	Fat.	Carbo-hydrate	Ash.	Calories
White bread	35.4	9.5	1.2	52.8	1.1	75
Graham	32.3	8.5	1.8	55.9	1.5	80
Rye	31.8	10.1	.7	55.9	1.5	76
Biscuit	22.9	9.3	13.7	1.5		108
Coffee Cake	30.1	8.6	6.6	58.9	.8	90
Drop "	16.6	7.6	14.7	60.3	.8	117
Sponge "	11.6	6.5	9.6	70.3	2	114
Butter crackers	6.9	9.2	13.6	69.4	.9	127
Graham "	5	9.8	13.6	69.7	1.9	128
Oat meal "	4.9	10.4	13.7	69.6	1.4	129
Oyster "	4.3	11.	8.8	74.2	1.7	122
Soda "	8.	10.3	9.4	70.5	1.8	119
Doughnuts	17.9	6.6	21.9	52.6	1.	126

Apple pie	43.2	3.3	9.8	41.7	2.	78
Custard	62.4	4.2	6.3	26.1	1.	52
Tapioca pudding	61.8	3.6	3.7	30.	.9	49

It will be well to remember that the composition of bread, crackers and pastry vary greatly, according to the amount of butter, lard, sugar, eggs, milk and other ingredients that may be added.

VEGETABLES.

Artichokes	79.5	2.6	.2	16.7	1.	23
Asparagus	94.	1.8	.2	3.3	.7	7
Beans, dried	13.2	22.3	1.8	59.1	3.6	99
" Lima	11.1	15.9	1.8	67.	4.1	101
" string	87.3	2.2	.4	9.4	.7	13
Beets	87.6	1.1	.1	9.6	1.1	13
Cabbage	90.3	2.1	.4	5.8	1.4	10
Carrots	88.2	1.1	.4	9.2	1.1	13
Cauliflower	90.8	1.6	.8	6.	.8	11
Celery	94.4	1.4	.1	3.	1.1	5
Green corn	81.3	2.8	1.1	14.1	.7	22
Cucumber	96.	.8	.2	2.5	.5	4
Egg plant	92.9	1.2	.3	5.	.5	8
Greens	82.9	3.8	.9	8.9	3.5	17
Kohlrabi	91.1	2.	.1	5.5	1.3	9
Leeks	91.8	1.2	.5	5.8	.7	9
Lentiles	10.7	26.	1.5	58.6	3.2	102
Lettuce	94.	1.3	.4	3.3	1.	7
Okra	87.4	2.	.4	9.5	.7	14
Onions	87.3	1.7	.4	9.9	.7	15
Parsnips	79.9	1.7	.6	16.1	1.7	22
Peas	10.8	24.1	1.1	61.5	2.2	102
Peas, green	78.1	4.4	.5	16.1	.9	25
Pickles	89.	.5	.5	5.4	4.6	8
Potatoes, boiled	73.7	2.7	.2	22.3	1.4	30

Potatoes, raw	78.9	2.1	.1	18.	.9	24
" sweet	69.3	1.8	.7	27.1	1.1	35
Pumpkins	93.1	1.	.1	5.2	.6	7
Radishes	90.8	1.4	.1	6.6	1.1	10
Rhubarb	94.4	.6	.7	3.6	.7	7
Ruta-bagas	88.7	1.3	.2	8.5	1.1	12
Sour krout	86.3	1.5	.8	4.4	7.	9
Spinach	92.4	2.1	.5	3.1	1.9	7
Squash	86.5	1.6	.6	10.4	.9	15
Tomatoes	94.4	.8	.4	3.9	.5	7
Turnips	88.9	1.4	.2	8.7	.8	12

Beets, potatoes, cucumbers lose from 15 to 20% in peeling.

Turnips, radishes, ruta-bagas lose 30% in peeling.

Rhubarb, 40%.   In peas and beans 50% loss in pods.

These tables do not clearly indicate the food value of the various vegetables.   It would appear that pickles have one-fifth the food value of potatoes, while in fact the nutriment in them is not readily available, and they have practically no food value at all, as they are not sufficiently soluble to be of use for waste.   Nearly all the green vegetables, with the exception of potatoes, have a large amount of indigestible fiber, but experiments have not been sufficiently extensive to give reliable percentages as to how much indigestible waste the various vegetables contain, but it may be assumed that celery contains from two to three per cent; turnips and onions, exclusive of peel, two per cent; cabbage and beets, three per cent; carrots and artichokes, four per cent.; green and string beans, four to five per cent.

The husk or bran (not pods) of peas and beans amount to about five per cent. and correspond to the bran envelope of wheat, but as the legumes are from ten to twenty times as rich as most of the green vegetables, the indigestible part is relatively small.

FRUIT.	Water.	Protein.	Fat.	Carbo-hydrate.	Acid	Ash.	Calories
Apples	82.	.5	.5	16.6	1.2	.4	21
Apricots	83.	1.1		13.4	1.2	.5	17
Bananas (yel.)	71.1	1.2	.8	22.9		1,	30
Black berries	88.9	.9	2.1	7.5	1.2	.6	16
Cherries	86.1	1.1	.8	11.4	.9	.6	17
Cranberries	89.	.5	.7	10.1		.2	14
Currant	84.7	.5		11.	2.15	.7	17
Figs	79.1	1.5		17.4	1.4	.6	24
Grapes	78.8	1.3	1.7	16.2	1.	.5	27
Gooseberries	86.	.4		4.6	1.5		
Lemons	89.3	1,	.9	8.3		.5	13
Muskmelons	89.5	.6		4.6		.6	49
Nectarines	82.9	.9	.6	15.9		.6	19
Oranges	88.3	.8	.6	7.	2.44	.6	14
Peaches	84.5	.5		14.2	.9	.8	19
Pears	83.9	.6	.8	14.2	.2	.5	19
Pine apples	89.3	.4	.3	9.7		.3	12
Plums	79.	.5		18.5	1.50	.5	24
Prunes	80.2	.8		18.5		.5	22
Raspberries	85.8	1,		12.6	1.38	.6	16
Strawberries	90.9	1.	.7	6.8	1.5	.6	11
Watermelons	92.9	.3	.1	6.5		.2	8
Whortleberries	82.4	.7	3.	13.5		.4	24
DRIED FRUITS.							
Apples	36.2	1.4	3.0	57.6		1.8	78
Apricots	32.4	2.9		63.3		1.4	78
Currants, Zante	27.9	1.2	3.	65.7		2.2	86
Dates	20.8	2.2	5.1	70.4		1.5	97
Figs	22.5	5.1		70.		2.4	87
Grapes	34.8	2.9	.6	60.5		1.2	79

Prunes	26.4	2.4	.8	68.9	1.5	85
Raisins	14.	2.5	4.7	74.7	4.1	102
NUTS.						
Chestnuts,fresh	38.5	6.9	8.	44.9	1.7	71
Peanuts	9.2	25.8	38.6	24.4	2.	160
Cocoanuts	46.6	5.5	35.7	11.	1.	115
Filberts, fresh	48.	8.4	28.5	13.6	1.5	
Walnuts	45.5	12.5	31.6	9.9	1.7	

The amount of sugar and acid varies greatly even in the same variety of fruit. The food value of fruits is mainly dependent upon the gum called pectose, and fruit sugar it contains. It is to be regretted that our present method of analyzing fruits, does not give satisfactory results as to the acids they contain. The very sour fruits, like the lemon and lime, have practically no food value for either fuel or tissue, but very great value for their acids. They must be regarded as cleansing agents. Most all fruits are more or less so.

PART II.

CHAPTER XXIV.

IDIOSYNCRASIES.

Idiosyncrasy is a peculiarity, in which one person is in some way affected in a different manner, under the same conditions, from ordinary people. It is applied to foods when there is a great dislike to some particular food, or where some particular food exerts an effect entirely foreign to what it usually produces. Idiosyncrasia is the term applied to peculiarities of smell. Both are closely related in their effects, and have not been given sufficient attention in their relation to health and disease. We are led to do so, because many people believe that individual peculiarities are so great that knowledge of food is of little or no use. If any of our readers take such a view, we have a troublesome question to ask: What makes the peculiarities? At first thought most people will say that they are "born that way;" but suppose we go farther and ask why people are born with idiosyncrasies? Do they come from some unknown realm, or are they transmitted characteristics? Here is the real key. Transmitted peculiarities were at some time acquired, and every one knows that acquired peculiarities are mainly due to habits or education. Who can doubt that if an American child a few months old, was taken to the heart of China, and reared as a Chinaman, but what it would eat substantially the same foods as the Chinese? This fact has so often been illustrated by taking children from civilization to barbarism, and barbarism to civilization, that it strongly tends to disprove the belief that people are "born that way." Take an illustration from the lower animals. A Texas cow or Texas pony that never saw corn will not eat it when first offered, but can easily be trained to do so.

Idiosyncrasies are either mainly acquired by habit or are the heritage of ancestral habits. A small per cent. are doubtless due to some strong mental impressions made upon the individual or upon the mother while carrying her unborn child. A careful study of the subject leads us to believe that idiosyncrasies toward food might properly be divided into three classes:

1. Those that are physiological.
2. Those that are due to habit.
3. Those that are due to mental impressions.

It must not be understood that individuals always manifest either of these independent of the others; for doubtless many have peculiarities about what they eat, which may be due to either or all the causes mentioned. The physiological idiosyncrasies are due to inability to digest certain foods, so that, as a matter of fact, most idiosyncrasies of which we take notice, are not idiosyncrasies at all, but irregular physiological action. Upon this we predicate the declaration, that one food will agree with one person as well as with another person under the same conditions. This sweeps away the notion that people's peculiarities make it useless to study food. It really does more; it proves the great importance of such study, because when we know why foods disagree, and the properties of foods, we will know why they agree at one time and disagree at another. The stomach that secretes but little acid, will poorly tolerate large quantities of lean meat, and such a diet will produce a feeling of weight in the stomach entirely independent of any gaseous fermentation. Those

who have an excess of acid will be distressed when they
eat starchy foods, especially bread, potatoes, beans, etc.
Such persons say they cannot live without meat, and when
they do not have it they always feel hungry, for the reason that they cannot digest starch. This is the most serious indigestion. Some physicians confound acid secretion
with acid fermentation. Sour stomachs and heartburn are
most common where the gastric secretions are weak and
do not call for a meat diet, as many suppose, but an aseptic (not readily fermentable) one. Lack of ability to digest
certain foods, indicates physical abnormality. Examples
might be multiplied wherein various foods agree or disagree, depending upon the needs of the system, the activity
of the stomach, and the condition of the intestines, pancreas and liver. It may also depend upon the blood at
the time the food was eaten. If the blood be laden with
effete matter and poor in quality, because of a previous
unsuitable diet, the general tone of the digestive organs
will be impaired. All of these are factors which make it
difficult to determine what agrees and what disagrees. And
people are often mistaken about their supposed peculiarities, but as the incompatibilities of foods and weakness of
digestion have been previously discussed, the idiosyncrasies due to habit will be most interesting. Nothing more
strongly illustrates the effect of habit than the universal
fondness for foods "like mother used to make." This is
one of the strongest traits in human character and emphasizes the extraordinary importance of proper home training. The habits of early life seem to be interwoven with
every fiber of our existence, and while there is no one in
this world so revered as she who gave us birth; no name
so dear as that of mother; yet it is distressing to realize
that disease and death-producing habits are not less destructive because made familiar to us by her hands. When

mothers realize their obligations to their children, they will not cultivate appetites and tastes for foods that are in effect the same as murder. Most idiosyncrasies of habit are due to ignorance about foods. Could anything better illustrate this than some of the ridiculous notions people have about what they eat? Let us consider a few of them:

Oatmeal.

Some people say that oatmeal sticks to the stomach; others, that it is too heavy and unfit for food in hot weather; while still others declare that long cooking makes it "slimy" and not fit to eat.

Bread.

Most Americans think that no bread is fit to eat, except fresh, doughy bread, loaf or biscuit.

Tea and Coffee.

That only excessive whisky drinkers are inebriates, and that tea and coffee are good nerve tonics and strengthen the system.

Celery.

That it is a brain and nerve food.

Soup.

That it is particularly wholesome and nourishing.

Pickles.

That they are eaten by people when in love and are good for young girls and "old girls."

Prunes.

That they are very laxative or cause diarrhoea.

Tomatoes.

That they cause cancer and are good food for children.

Popcorn.

That it is wholesome as ordinarily eaten.

Fried Meats.

That they are fit to eat.

Radishes.

That they aid digestion and act on the liver.

Ice Drinks at Meals.

That they cool the system and aid digestion.

Condiments.

That they are beneficial.

Green Corn.

That it is healthful for human beings and bad for swine.

Alcohol and Beer.

That it increases strength and adds to the general well-being of the imbiber.

What can be expected of people who are governed by such expansive ignorance?

By way of parenthesis and confidential advice to dyspeptics, we might add, that if their attacks do not come often enough they should eat lobster salad, ice cream and rich cake between 10 p. m. and I a. m.   If you expect the arrival of your family physician, a good meal of cucumbers, vinegar, milk and ice water, will very likely make you glad to see him.   If habits were not so pernicious and far-reaching in their effects, idiosyncrasies would not be worth considering; but as many people's lives are spent running from or running after their idiosyncrasies, it is time that attention should be given to the causes which wreck so many lives.   Parents should first purge themselves of their suicidal habits and then start their children right.   A generation or two ago, when disease-breeding luxuries were not so easily obtained, children were reared with a view of becoming strong, able-bodied men and women, who could assist in building homes; now, children to a great extent command the obedience of their parents in all their whims and follies.   Parents are the guardians of posterity, and no language is strong enough to portray the misery which results from improper feeding.   A diet mainly composed of sweet-meats and highly-spiced foods so perverts the nerves of taste, that plain, wholesome food is too tasteless or disagreeable. The result is that sooner or later the effects of such habits bring disease, and then they bewail their misfortune as a curse from God or Satan, whereas, it is the curse of personal and parental folly. The first step towards reform is sensible cooking; and then parents must see that their children eat a suitable variety of foods. We have often seen children make a meal on canned tomatoes, also on fried eggs and fried meat, corned beef, bananas, green corn, raspberries, pickles and cake. Children are often allowed to pepper their food until black and then cover with strong mustard. The effects of allowing children to have what they wished, as mentioned, varied from "unwell" to death. The greatest obstacles to health is pampered appetites. People will not eat what they do not like, no matter what the consequences. Some people urge this as an objection to the study of foods; but it really only emphasizes its importance. When will people like what will keep them healthy and strong? The answer is simple: It will be when they are taught to eat wholesome foods in their childhood. Here is the strong point we make: People cannot form correct habits as to what and how to eat until they know the properties of foods and how to prepare them. Disease and death have been accepted as the inevitable, with barely a thought about individual responsibility. Children are sent to school that they may be trained for the duties of life, but the most important thing is neglected, or taught in a way to be of very little practical benefit. What is the use to train the mind while the body is being killed? The proper use of food must in time hold the highest place in education, both in the home and at school. The value of moral and religious training is partly realized, but the religious world is in almost total darkness about the relation between the physical and the moral life. Morality cannot be high when the tone of the body is low. Immorality is largely due to physical or nervous propensities that are either inherited or due to the violation of Nature's laws.

If reformers will give two-thirds of their labor to teaching people the way to a perfect physical life, the other third will have ten times the results towards the moral and spiritual regeneration of the race.

The third class of idiosyncrasies are not so common, but much more difficult to overcome. These are both pre-natal and post-natal impressions. The action of the human mind is one of the most unexplainable things of nature. Its freaks and variations are unlimited, and are to be observed outside, as well as inside of lunatic asylums. The influence of the mind over the body is so great that a very enthusiastic religious order undertake to heal all diseases by faith, which is, in fact, a mind cure. Some physicians estimate that forty per cent. of the ordinary diseases exist only in the imagination, and the success of the faith healers and hypnotists would seem to give strong support to the view that a large per cent. have no other existence. It is reported that there is an institution in Paris where sham surgical operations are performed on those who think that nothing else will cure them. It is said that the patient is given an anaesthetic, a scratch and a few stitches, and is then cured. Very amusing instances of how the mind is affected are reported from time to time, and as an apt illustration of what imagination will do, let me report what the doctor declares actually happened:

"One evening, about seven o'clock, I received a telephone call to come to X.'s at once; that he had a fish

bone in his throat and was about to die. I immediately
gathered what instruments I thought might be needed
and hastened to X.'s residence, and found him lying flat
on the floor, writhing in agony and blue from holding his
breath, because of his fear that breathing would draw the
bone further down his throat. The family were wailing
and hysterical, and were under the belief that the head of
the household was about to pass over. Upon inquiry I
found that the patient had eaten fish and dry toast, and
after a careful examination, nothing was discoverable except a slight scratch on one of his tonsils, probably made
by the toast. After a moment's reflection, I concluded
that it was necessary to relieve the patient's mind, so I
told him I would remove the bone in a moment; and, under pretext of sterilizing my instruments, went to the
kitchen and got a fish bone, fixed it in the instrument
and went through the motions as if to remove the bone,
pricked the tonsil slightly, withdrew the instrument and
held up the supposed offender. The effect was magical;
and, after looking at the bone and taking two or three
swallows of water to see that all was clear, he declared
that the relief he experienced was something remarkable."

Physicians often cure by suggestion, as illustrated by
the experience of another physician. One of his female
patients had a chill every day at eleven o'clock. After
treating her for several days without apparent benefit, he
concluded that it was more hysterical than anything else;
so he told the patient that he would have to give her an
extraordinary remedy; that it was very dangerous if not
used just right, but that it could not possibly fail to stop
the chill. The doctor then gave her a small bottle of
water and instructed her to take it exactly ten minutes before eleven—the time the chill usually began. The doctor
took particular pains to impress the patient with the fact

that she could not possibly have another chill—and she didn't.

Repugnance to certain foods is often due to mental impressions. We know a cultured gentleman, who is nauseated at the sight of a raw oyster. He explains his peculiarity in this way: When he was about ten years old he made his first trip to a seacoast town where oysters were plentiful. Now, to an unsophisticated country youth, a raw oyster is certainly a slimy, repulsive looking object, and at first sight, to see a man gulp down big, slimy-looking oysters, made such a nauseating impression on him that he cannot to this day tolerate raw oysters, although the incident occurred more than fifty years ago.

One of the most noticeable effects of pre-natal influences is a man who constantly appears as if drunk, although he does not use alcoholic liquor at all. He has ordinary intelligence and physical vigor. His peculiar condition is due to his mother's fright during gestation, at some threatened danger from a drunken man.

A frequently-observed physiological perversion of the appetite is that which so often occurs during the period of gestation. During this time there is often a vague longing for some special article of food to which the person may or may not have been accustomed, e. g., ice cream, etc., and occasionally this desire extends to unwholesome articles of food, as decaying fruit, and even to substances not used for food, e. g., coffee grounds and even dirt.

These perversions may be very distressing, especially it the particular article desired cannot be obtained.

In contrast with the perverted desire, we have aversion to certain foods which previous to the advent of gestation had been liked. These likes and dislikes, to some degree, may be transmitted to the offspring. On account of the possible effect upon the child, the mother should, so far as possible, avoid anxiety concerning any peculiar tendencies of the appetite which may arise, and confine herself to the usual diet to which she is accustomed, if it be wholesome.

A discussion of pre-natal influences would make a volume of itself, and a good understanding of the subject is necessary for race-elevation. The object of all that has here been presented is to show that we are largely creatures of whims and accident, and that the mind can be trained to govern appetite and throw off disease. Very strong mental impressions can usually only be removed by hypnotic suggestion, but ordinarily our likes and dislikes can be controlled by the will. The appetite for celery is acquired, as very few grown people will eat celery the first time it is offered. People are certain to dislike whatever they make up their minds that they won't like, and can easily learn to like anything the same way. We know a boy who was given pills in peach preserves, and it was years before he could disassociate them. It is very nonsensical to cultivate a dislike for foods because their appearance does not always please our vision.

An unnatural or perverted sense of smell is not so common as that of taste, but it is more annoying. The causes are much the same as idiosyncrasy, except that disease is more of a factor. Errors in diet often cause nausea, vomiting and fainting, which, in turn, may affect the sense of smell and create a disgust for food, and increase the illness for the want of it. Strong extracts of cologne will often bring on attacks of dizziness and vertigo, and the odor of flowers frequently excite violent attacks of hay fever. The odor of the oleander has been known to create a disgust for flowers that never could be overcome. The ancients well knew the influence of flowers, and they were used as a potent aid in love, intrigue and even crime. Tube roses have been known to produce melancholy to a degree of wild insanity, and it can be truly said of them that they are hardly less sad than beautiful. Almost every person dislikes one or more foods, because of the odor produced. When the smell of any particular food or all foods give offense, the odor arising from cooking should carefully be avoided, and if necessary the meals should be eaten in the open air. A great deal can be accomplished by firmness. Most horses will not eat pumkin until starved to it. The method is very simple; the horse is put in the stable and pumkins in the feed-box. No other food or drink is given until the pumkins are eaten, and ever afterward the pumpkins are as well relished as oats. We are not prepared to advocate this method as an aid to the correction of a perverted sense of taste or smell, but we do know that hunger makes great changes in our tolerance of food. This is illustrated by a gentleman who took a long bicycle ride in the country. For some unaccountable reason he had great antipathy to the odor of raspberries and could not bear the sight of them, for he in some way associated them with the odor of a certain little animal known to be capable of nauseating an entire neighborhood. About ten o'clock in the morning he became ravenously hungry, so he decided to stop at a farm house and get a lunch; but to his chagrin he found no one there except a little girl. She told her uninvited guest that she did not belong to the family and could not give him anything to eat except the berries which she had gathered to which he was welcome. Not thinking of raspberries, and being hungry enough to eat anything, the kind offer was accepted. The little girl brought the berries, and before he could refuse, she had delivered them into the gentleman's hands. The

novelty of the situation, together with his hunger, over-
came his repugnance to the fruit, and he has been fond of
them ever since.

Judicious flavoring and repeated trials will bring the
appetite to favor suitable foods.

The best cure for all peculiarities is to prevent them,
by proper living, as we have pointed out.

CHAPTER XXV.

CAUSES AND SIGNIFICANCE PAIN.

Pain is the cry of an injured nerve. It often masks its real intention and plays hide and seek in the human body and should therefore be studied with great care, so that its characteristics may be fully understood. Most people understand it to be a gauge or measure which indicates the violence of the disease which causes it, but the danger from the disease is frequently in an inverse ratio to its intensity, so that the absence of pronounced pain may be of much greater significance than its presence, and while it indicates discordance in the human mechanism, it may herald the exact locality in which there is trouble, or may be very remote from the parts affected.

Many persons suppose that serious diseases always produce pain, and lack of knowledge on this subject has made thousands of persons victims of fatal, insidious diseases, that send no advance agent—pain—to herald their approach. These painless destroyers of life glide into our very vitals, and like the poisonous reptile in our pathway, strike us a fatal blow without even a pain to warn us of their existence.

This would seem to make the value of pain in the diagnosis of disease somewhat uncertain, and yet careful study of its intensity, recurrence, location, and even its absence, greatly aids in the diagnosis of disease. In order to determine the significance of pain, we must take into consideration the relative sensitiveness of different individuals. We have all observed instances in which persons have complained a great deal from comparatively slight pain.

while others have endured severe pain with little or no
complaint. If a patient says that he has a severe pain, we
must determine by the appearance and character of the
person, what this actually means. Does it mean that he
is suffering from some serious disturbance, or is he simply
exaggerating, perhaps unconsciously, what to some others
would be considered slight pain? Again, it will some
times be observed that children, and often adults, will
endure pain without admitting it, if they fear the treatment that might follow, especially if that treatment be
surgical. On the other hand, persons will often complain of pain and suffering, in order to attain certain ends.
Examples: persons have often been known to complain
of pain in order to secure morphine or some similar drug
which they crave.

Why is it that some people actually feel pain more keenly
than others? This may be accounted for on several
grounds. Some persons naturally have a sensitive nervous system, and any irritation will cause greater excitement of their nerves, than it would in a person who
naturally has a sluggish nervous system. As a rule, persons inured to hardships can stand more pain than those
who have always had an easy life. It is for this reason
that those who have had much pain can endure it better
than one who has always been free from it. It is fortunate
that pain is the forerunner (so-called) of many diseases,
and it would be better perhaps if it were present in the
beginning of others; for it is the one thing which by its
persistent annoyance will drive its possessor to determine
its cause and seek relief. It is not uncommon to learn of
persons who have neglected some serious disease because
it had not caused them much pain. Alas, we would that
pain were always the forerunner of disease; for too often
it comes too late and the victim is shocked to learn that

an apparently slight trouble is of a serious nature. It consequently happens that a simple distress, or uneasy feeling, develops sooner or later into actual pain. Now the time to seek relief is soon after the distress is felt, and not wait weeks or months for it to disappear, without removing its cause. Only a physician can interpret the real value and importance of a given pain, for the real cause does not always lie immediately in the region where the pain is felt, but may be quite remote.

Under this head comes what are termed reflex pains. Example: heart disease will sometimes cause pain in the back and left arm; liver disease may cause pain in the right shoulder and back; spinal disease, pain in the legs, etc. How often we hear people who are suffering pain, say, "Oh, I think it is only a little rheumatic trouble due to the weather." While this is sometimes in a sense true, it is also true that some serious diseases early cause pains which may readily be mistaken for some slight trouble.

Let us not forget that pain under all circumstances, indicates illness somewhere, and it might be almost truly said, that aside from the pain of child bed, there are no natural pains. The belief entertained by some, that children have "growing pains" is, to say the least, a great mistake, for neither slow nor rapid growth produces pain, and it is no part of healthy development. No perfectly well person feels pain, and when he does so, it should invariably signify to him that some part of his physical mechanism is out of order, and in need of attention. It may be compared to the pounding or squeaking of an engine, or any machine, which should run smoothly; when it is heard the engineer knows that something is wrong, and at once tries to locate the cause and remove it. If it is necessary to take such precaution, with an inanimate machine, why should we not take equal or greater precaution with our living mechanism, since our very existence, as rational, active beings, depends upon its preservation? Knowing all this, why do we not more often heed the kind warning of nature, remove the cause of an injury, and repair the part to which pain directs us.

The savage, inured to pain from early infancy, meets it with a stoicism which is worthy of admiration, but which science and experience teaches, is too often the result of ignorance and superstition. Civilized men should not tamper with, or tempt disease, by violating the inexorable laws of nature. If we heed her first warning, she may be lenient, but if we disregard them, she may be cruel in the extreme. The saddest information which the physician gives, is, "you have neglected your trouble too long," and the equally sad reply of the patient, "I did not think it was so bad, as I have not suffered much pain."

These words may seem trifling to the reader, but stop one moment and reflect. May they ever apply to you? Yes, they may at any time apply to any one, and especially to those who disregard the rules of good living, and the warning of slight pains. A careless engineer who disregards slight defects in his machine may soon find that it has suddenly stopped—possibly capable of repair but often beyond such a possibility. If we could be so impressed with these facts as to be impelled by them, there would be fewer cases of premature age and early death; fewer cases of impaired and useless human machines. Whence arises that ill-sounding and never acceptable word, *incurable?*

Too often it results from repeatedly disregarded pain, until the disease has advanced too far—nature has too long been outraged. In those cases where there is slight pain with the onset of some insidious disease, as chronic kidney, heart, or lung disease, etc., we can understand how a person who does not recognize the importance of a slight pain, or even physical discomfort, may compromise himself to disease, but to one who does know the significance of pain, it is gross carelessness. The location of pain is also a point worthy of consideration, for it is often misleading. The real source of disease may be far from the place where the pain is felt. To illustrate, let us consider some common examples: headache—one of the most common pains—may result from disturbance of the stomach, or bowels, or it may arise from some difficulty of the nervous system, or again, from actual disease of the brain. What is so commonly called heart-burn results from disturbance of the stomach, and is in no way a heart pain, simply felt in the region of the heart.

Diseases of the female organs usually produce pain in the back, and not at the seat of the disease. Pains in the intestines can often be recognized by the patient, because of its colicky nature, but pains of other abdominal organs are not so easily recognized. How often people complain of pain in the kidneys, when in reality it is not in the kidneys, but in the muscles of the back, and may be some distance from the kidneys. Pain in the kidneys is not one-half as common as many suppose. The character of the pain is as important as the location. In the first place, sharp pains more often indicate acute diseases, and dull, aching pains, chronic diseases. Of course, this rule has exceptions, which only emphasize the importance of finding out the cause of pain whatever its character.

The intensity of pain also has considerable significance in disease. A pain does not have to be very intense in order to indicate a serious trouble, for some grave diseases, often exist with only a slight, if any pain at all. On the other hand the most intense pain is often found in acute and relatively slight diseases. Acute indigestion is a comparatively slight disease, as regards the probability of recovery, but the pain is often intense, while a serious form of dyspepsia may show but slight pain. Finally, we should consider the duration of pain. If pain persists for hours or days it is usually of a more serious import than if it is transient.

Neuralgic or rheumatic pains are often transient, while the pains of organic diseases are often protracted, but may vary in intensity or stop temporarily. It is usually for persistent pains that people seek relief. They can stand pain part of the time, but they cannot as a rule stand it all the time. Now with what has been said about pain it becomes readily apparent that the individual should not trust too much to himself for its interpretation, but if it is at all constant, he should secure the opinion of some physician who can tell him whether it is of any real concern. This might well be illustrated by mistakes which are quite often made in regard to pain in the abdomen; not uncommonly pain in this region has been thought that of colic or indigestion, while it proved to be the pain of appendicitis. In the first condition there is no great anxiety, but much in the latter and it should be recognized as early as possible, since a great deal depends upon the manner of treatment.

Let no one think slightly of pain, for if heeded in time serious illness can often be prevented, and much suffering avoided. The doctor, skilled in examining patients, learns to distinguish the importance of the pain, and can often relieve a patient's anxiety or inform him of its true significance. People must not think that they can persistently disregard natural laws, until they suffer from organic disease, and then expect a physician to undo all their past mistakes. The best time to treat disease is before it exists, and then the treatment is very simple, inasmuch as it only means correct living. The penalties of unhygienic living are often long delayed, but they are almost sure to come sooner or later. Let us guard ourselves, lest we suffer from having abused nature.

CHAPTER XXVI.

FEEDING THE SICK.

It was formerly supposed that drugs cured disease, and as a consequence, very little attention was paid to feeding sick people. The modern physician strives to aid nature. Nearly a half century ago, a noted medical writer stated that of all the means known for the cure of disease, none was so powerful as a proper adaptation of food and drink. It is to be regretted, that the importance of feeding, has not been more generally recognized, and this being true, it is no wonder that methods of feeding should escape attention. The first requisite is suitable food, and the nurse must see that it is properly prepared. It should be the duty of some one in every household to take charge of the preparation and feeding in illness. It is not enough for the physician to give directions as to the food. The nurse must know that the milk is sweet and pure; that the broths are properly made; that the toast is thoroughly dried and browned; that the gruels are thoroughly cooked and that the fruits are neither green nor over-ripe.

The patient's appetite and peculiarities must be watched and it is of the greatest importance to find out his likes and dislikes, and how to flavor and serve what is agreeable. Study the patient's whims and agree with them, and under no circumstances should the patient be directly antagonized. If necessary to do things radically different from what he desires, agree with him in speech, but do what is best. If the patient wanted something very harmful that would require some tact. If not amenable to reason and the condition serious, it may be well to say

that he can have all he wants; that the doctor directed that
he be compelled to eat ten or twenty pounds. If per-
suaded that he must eat a disagreeable amount, it often
happens that they will not even taste the article. No very
specific directions can be given. The patient's mind must
be appeased as well as his body fed. There may be no
appetite and all food refused, but this may result from
offering unpalatable food, or due to some offensive way
in which it is served. As a general rule, patients should
be fed regularly. If the patient sleeps a good deal he may
be roused up; but if not, the best time to offer food is just
after the patient has slept, but never immediately after
severe attacks of pain, unless unavoidable. One of the
most essential things in good feeding is the patient's comfort. It is of the utmost importance that the patient be
made comfortable and able to receive the food without
exertion, otherwise he may dread the sight of it. Food
can always be made most appetizing to the patient by being served in an attractive way. This means clean hands,
clean apparel, and the best china in the house. Sick people are often much more observing than when well, and
great care must be taken not to present the same appearance when offering food as when doing chamber work,
otherwise the patient may associate the two and be nauseated at the sight of food. Hot food should be served
quite hot, and cold ones sufficiently cold to be pleasant
to the taste. It must be remembered that the sick are
more or less sensitive and whimsical, and great bulk is
particularly repugnant to a weak appetite. The practice
of leaving medicine bottles and remnants of a meal on a
chair or table, where the patient can constantly see them,
is very careless, to say the least, and calculated to make
the patient loathe the sight of food, and instead of a con-

stant vision of nauseating medicine bottles, bright, fragrant flowers will exert a beneficial influence.

The frequency and quantity of food to be given, depends largely on the condition of the patient. The digestive secretions will usually be the weakest during high fever. Patients are seldom fed at closer intervals than two hours, or farther apart than four or five hours.

If food causes nausea and disgust, it only does harm to offer it, no matter who advises it; but it does not follow that all foods will do so; and when one disagrees, something else must be substituted for a time, even if less suitable, until the patient can tolerate a better food. As a general rule, food should be given at regular intervals and in small quantities. Always carry small quantities of food to the bedside, and when the patient has little or no appetite, it is not advisable to ask what would be agreeable. If the patient be nervous and suffer greatly from pain, and therefore unable to sleep, he should not ordinarily be aroused for feeding; but if he sleeps much and is easily aroused, he may be fed at the regular periods. If he should insist on having some food, of doubtful use, prepare it without fat of any kind. If it be a solid, grind to a powder, as fine as flour, if possible; but it is usually better to give only liquid food, and if vegetables, they should be stewed and only the broth given; and if fruits, only the juice. In such cases, give teaspoonfuls and watch results. In giving meat broths, the oil floating on the cup should always be skimmed off with a piece of bread, before offering it. The care of the patient's mouth is hardly less important than the feeding, because a bad mouth may indirectly be the cause of death. The mouth is affected by fever, medicine, and foul secretions, which are likely to make it very uncomfortable and sometimes very sore; and in either case, it may destroy the patient's appetite. Now, many patients die of exhaustion, that would probably not have done so, had they been properly nourished; and this, in turn, may have been because of the condition of the patient's mouth. There are only two things to be done, and that is to cleanse and disinfect after each feeding. Use warm water, to which a little of some mild disinfectant, such as boracic acid, has been added, and then rinse with plain water. Soft brushes or swabs should be used where possible. Of course, in washing the mouth nothing should be swallowed. Chewing a slice of lemon has a remarkable effect in cleansing a foul tongue, and for this purpose probably there is nothing as serviceable. The lips should be moistened with salt water, or vaseline, or nut oils may be applied. Unconscious patients must not be fed anything but liquid food, and that through a catheter. This is necessary, because it is difficult to get food to the stomach in any other way, for the patient will not swallow. If the mouth cannot be opened and there are no teeth missing, through which the tube can be passed, then the tube must always be passed through the nostril. In the absence of any indications to the contrary, patients may be allowed all the cold water they wish. This is especially true in fever. During a chill, or where it is necessary to reduce the volume of blood, as in some disease of the heart, or puerperal eclampsia (spasms after child-birth), specific directions as to amount of liquid to be allowed, must be given by the attending physician. The matter of ventilation and sanitation do not belong to this book, except as an incident affecting the patient's appetite.

Many people in ordinary health are almost as afraid of "drafts" as of small-pox. It is, therefore, not surprising that sick chambers are often kept without pure air. The sick room must be kept sweet. No sick person can have an appetite or relish foods when kept in a foul, stifling atmosphere. Warmth and fresh air are the first of all considerations, and air exerts the greatest influence in diseases of the lungs. All vessels used in the sick-room must be disinfected and cleaned with boiling water every time they are used.

Diet in Acute Diseases.

Many suppose that feeding in acute diseases is unimportant, because they are usually of short duration. This is a great error; for who knows the duration of any attack of illness? The patient should be kept in as good a condition as possible to resist disease, and to be able to more quickly resume the duties of life.

A few years ago, a large per cent. of typhoid patients died of exhaustion after three or four weeks' illness. Now it is possible to carry them the same length of time with very little loss of weight. Fever destroys tissue at a high rate. This calls for diet rich in proteids. There are some general principles which apply in nearly all acute diseases. Briefly enumerated, they are as follows:

(1)   Foods must be well cooked and easy to digest.

(2)   They should be given in fluid, puree, semi-solid or powdered form.

(3)   It is generally necessary to give small quantities with greater frequency than in health.

(4)   Foods should be given when the body temperature is the lowest.

(5)   All foods must be bland and unirritating.

(6)   No iced drinks should be given except by advice of attending physician.

The foods most commonly used in acute diseases, except water, are as follows:

Milk holds first place, and can probably be used in some form in all cases. Plain milk may first be tried, either cold or hot, then pasteurized, sterilized or peptonized.

If these do not agree, try milk and barley water, or milk and gelatine, or milk and "slippery elm" water. In diarrhoea use milk and lime water, and in constipation milk and soda water (see "Milk," for methods).

When desirable to use the largest quantity of food, the milk should be thickened with well-cooked starch, either rice, barley flour, arrowroot, sago or corn starch. The fine flour starches should be put to cook in cold water and gradually brought to a boil and kept boiling for an an hour or two. Coarse meals should be kept boiling from three to five hours and then strained. Meat broths are used extensively, but as ordinarily made they contain but little nutriment. Meat may be used in powdered form to better advantage. The most practical way to powder meat is to grind it at least six times, and each time it is run through the grinder, the solid or stringy part should be removed. This will reduce the bulk about one-half, but the refuse may be used for broths. The powder may be macerated in cold water and then gently warmed. If the patient is very feeble, it should be strained.

Eggs may be used with either broths, milk or alcohol, where the latter is prescribed it will usually be desirable to use it as a vehicle for food. They can best be used without cooking, when beaten or stirred into hot broths or milk. Skill in flavoring may save life, because recovery may depend on the strength of the patient, and that in turn on the amount of food that can be taken. Lemon, pineapple, vanilla, nutmeg, cinnamon and fruit flavors generally, may be employed. Some of the prepared infant and invalid foods may be used to great advantage. The use of fruits gives most concern, because of their sugar and acid, which may quickly ferment. Sometimes they seem to exert a restorative power that is marvelous. The sweet fruits seem to ferment too quickly, and the sour

ones are incompatible with milk, and sometimes with the
medicines administered. If attending physician fails to
give any directions about fruits, it would be well to ask
whether acids would be incompatible with any of the
drugs administered. The sour fruits should be given with
the egg or meat broths, and not with milk, and always free
from seeds and skins. The juices of stewed fruits should
be used for their flavor, rather than the nutriment they
contain. Great care must be exercised not to give fruits
that are tainted with decay, or that contain any solid or
tough substance. Nothing but the juice should be given,
with the possible exception of mellow peaches, baked
apples and banana meal.

It is certainly advisable to give daily all the food that
can be digested, whether that be a pint or three quarts.
We doubt the propriety of giving more than three pints
or two quarts of milk per day in typhoid cases. More
nourishment will be needed, but meat-powder or beaten
eggs, with an increased supply of water, will bring best
results. It is best to add barley water, well-cooked starch
or gelatine to milk to prevent the formation of large
curds. There are no inflexible rules—feeding must be
adapted to the patient.

Drinks.

Not many generations ago, the sick died of thirst, because the people were so ignorant as to believe that water
was a strong ally to fever. Now water is administered
inside and outside, and fever is controlled by it when all
other remedies are impotent. Patients who are rational
will ask for water; but those who are unconscious, should
be given water at regular intervals. If milk be used extensively, the need for water will be much less than if
smaller quantities of fluid be ingested. Sour lemonade
is one of the favorite drinks in fever. If there is any

tendency to sour stomach, no sugar should be used. It must not be given with milk or starch gruel, nor with incompatible drugs. Both tea and coffee should be refused, but cereal coffee may be allowed, and very useful by way of variety. Coffee may be used for flavoring milk or other foods. It must not be allowed to boil, and should be steeped only a short time.

If the patient likes chocolate, a little may be added to cereal coffee, but should be strained before using. Grape juice, unfermented fruit juices, and natural mineral waters are usually allowable, and sometimes very beneficial. Cold drinks must be slowly sipped, otherwise they may greatly interfere with digestion.

CHAPTER XXVII.

CAUSES OF INDIGESTION.

The diseases of the stomach and other digestive organs are so nearly universal in this country, and so closely related to each other, it may not be amiss to call attention to a few general facts as a prelude to a more specific discussion of the causes which produce them, and the maladies incident thereto. Primitive man lived much more in harmony with his stomach than our modern, so-called highly educated being. Civilization, with its inventions, has done much to elevate man and produce external comforts, most of which react on his physical nature; and this is especially true of his digestive organs. The stomach might aptly be called the boiler of the body; and when we think of what its owners compel it to endure, the question naturally arises, "Had man's stomach been constructed of aluminum, would he still have found some way to destroy it in the gratification of his perverted tastes?" As man now lives, his stomach is totally inadequate for the uses which he makes of it. Had it been constructed of some material that would not corrode, that would stretch to unlimited proportions and then set on springs, it might have met the requirements of modern usage. Modern man is a creature of boastful progress; but his very progress has brought him habits of self-destruction. Nature demands activity; whereas, the constant effort of man is to contrive some way to avoid work, and increase his luxuries—the highway robbers of health. Our savage ancestors were giants in strength and stature, and we, their dwarfed descendants, resemble them only as a shadow resembles its substance, and the best that can be said of us is, that we are a badly executed miniature, painted by the cramped, nerve-racked hand of modern civilization. There are several ways of using a candle. In olden times they lighted many candles, at one end, which made a bright light and burned long. Modern man is nothing if not ingenious, and seemingly economical. He lights both ends of his candle, saves candles and candelabra; but alas, how quickly burned out! We forget that pain and disease is the base alloy that makes our lives a counterfeit of Nature. If we would be free from physical infirmities or cure them, we must study their causes.

Causes of Disease.

The source of disease is sometimes obscure, but generally speaking, its causes may be divided into two general classes: (1) Those that come from extraneous sources, which are called contagious or infectious diseases. (2) Those that come from within, from poisonous products generated in the body, or some form of mal-nutrition. It is a great mistake to suppose that most of our ailments are unavoidable. A few of the more virulent diseases, such as diphtheria, are probably not dependent on the ill health of those whom it attacks for a starting point; but most germs have little or no effect on those who are in perfect health, while those who are already diseased, are easy victims. Some physicians say that nine-tenths of the ordinary diesases are caused by auto-intoxication— self-poisoning. This estimate may be too high, but all physicians of high attainments agree that a very large per cent. are so produced. To put it another way: we allow effete matter to accumulate in the system, or take substances into the body which form poisonous compounds, and disease, or at least weakness, results. Now, as good health is the armor of Nature, the system is defenseless without it. It is our purpose to show how disease originates by pointing out the most common errors in our habits of living, and the characteristics of the diseases produced. The immediate sources of contagious diseases are beyond the scope of this book, for they are all dependent upon bacteria or other organisms. We therefore pass them by. Modern investigation has thrown much light upon self-inflicted diseases, and how they originate. This has come from a better knowledge of chemistry and the use of the microscope. Our bodies are real laboratories. We eat food and it is converted into heat, muscle, nerve, fat and bone. The processes are many, and none can be safely omitted. This fact seems to be generally overlooked. We eat to live, but most people exist (not to live) to eat. There is no teamster so dull but that he knows there is a limit to the capacity of his vehicle; no miller that does not know that he cannot put two barrels of flour in one; but how many people have ever given any thought as to the capacity of their digestive organs or the needs of the system? In this respect they have far less regard for themselves than they have for any piece of machinery they possess. To do good work a machine must not be fed beyond its capacity, and it must be kept clean. Just so with the human body. If properly fed, and kept clean, free from effete matter, there will be no disease. A good many attribute their illness to overwork, or the weather—sometimes to la grippe or malaria. Old people who are rheumatic, gouty and stiff, are certain that it was the hard work done in early life that makes them so. Is it any wonder that the rising generation is not on good terms with work? There is an occasional person who gets sick from overwork; but the overwork that causes most sickness is that done at the table.   Very few people injure themselves by physical labor; but a small number do from mental work.   Overwork is usually a nice-sounding name for over-stimulation from tobacco, alcoholic liquors, tea or coffee, which disturb digestion and prevent the relaxation and rest that is essential to good health.   Injudicious diet, lack of exercise, and stimulants, "overwork" thousands of people.   Any well-nourished person can work nine to twelve hours a day without injury, but the people want to be told that it was overwork, rather than bad habits, that causes their illness.   Work is not a curse, but a blessing—though most people don't want to be blessed that way.   Some "overworked" people so seldom use their feet and legs to move themselves, if it were not for uncomfortable footware they would forget they had such useless appendages. "Nervous exhaustion" might often appropriately be named pedal inanimation.   The other supposedly great cause of disease—the weather—is also only a small factor.   And then only in connection with one of the real causes—the imperfect elimination of waste.

Waste of the Body.

The waste of the body is eliminated through the lungs, skin, kidneys and bowels; and whenever it is not promptly removed from the system, disease results.   The waste is made up of three elements:

1.   The dead tissue of the body.
2.   Indigestible particles of food.
3.   Excess of food taken into the system over and above its needs.

If the mere smell of decaying tissue-foods, such as meat and eggs, makes one sick, does it not follow that it would have a worse effect when in the system?   The effete tissue and excess of food, especially meat and eggs, are really poisonous.   People know something about the necessity of food, but seem totally unconcerned whether the waste is removed or not, although it is of vital importance. Every one knows that human life cannot exist without air, but they do not realize that it could not long exist if every pore in the skin were closed. A large amount of impurities is thrown out through the lungs, and foul breath (except it come from mouth or nose) is one of the best evidences of how the system tries to cleanse itself. A wet sponge will not absorb as much water as a dry one; nor will air, laden with impurities, carry away as much waste from the system as pure air. An active skin is almost as essential as pure air, and if generally recognized, disease would be far less common. All intelligent people understand the necessity for keeping the skin clean, but lose sight of the fact that it must be kept warm. This explains why the changes of the weather make people sick. A chilly or damp day may close the pores of the skin; and if the other outlets be inactive, a cold is the result. This is especially noticeable in those who eat more than their system needs. An excess or too little clothing (especially on the extremities), overheated rooms that dry the skin, are causes of cold and indigestion. The nitrogenous waste of the body is removed through the kidneys; but as they are mainly affected by errors in diet, nothing need be said about their action. Non-elimination of waste, on account of constipation, is so common that it demands separate treatment under diseases of intestines. No matter how we live, there will always be a certain amount of tissue that is being removed and entering the circulation. If the excretory outlets are insufficient, the poisonous matter is kept in the system, with results that vary from discomfort to death.

Lack of Exercise.

The necessity of labor for most people, gives sufficient

exercise; but many women and business men take too
little.   Those who do heavy work need a great deal of
food, because food is burned up in force-production.
Besides this, great muscular activity shakes the dormant
digestive organs into activity, and assists in the elimination
of waste.   The main difficulty is to provide the proper
amount of food for a certain amount of exercise.   People
who lead an active life will eat as much or more on Sunday, when they do nothing, as when they are at hard labor.   This brings us to the principal source of disease:

Lack of Adaptation in Diet.

Under ordinary conditions every organ in the body is
more than able to perform the functions for which it was
intended, and there should be no disease; but so long
as people utterly disregard the law of supply and demand,
in the matter of feeding themselves, so long will the human family be cursed with it.   The various organs of the
body are dependent upon each other, and all are dependent upon good blood, which can only come from food
adapted to the needs of each individual.   It should be
constantly borne in mind, that the ordinary diseases result either from poison or starvation; poison from dead
tissue or decay of foods; starvation, because the foods
did not furnish the essential elements of life, could not be
digested, or was too small in quantity.   It sounds paradoxical to say that one lacks nourishment, when already
consuming twice as much food as needed, but such is frequently the case.   It is not what we eat, but what we
digest and assimilate that sustains life; and there is no
fault that interferes with digestion more than an excess of
food.

Overeating.

Too much food unduly distends the stomach and weakens its contractile power, thus destroying, in a great meas-

ure, its activity—churning movements. This might be
illustrated by trying the strength of your arm when
stretched as far as you can reach. Besides this, there is
a limit to the amount of digestive secretions, and if these
be only sufficient to digest twenty ounces of food, it is
manifest that twice that quantity could not be digested.
Now, what results? The food will most likely remain in
the stomach too long and decay, which cannot be cor-
rected after the food passes out of the stomach. The
blood thus becomes filled with crude and often poisonous substances. This is what produces the languor, headache and general discomfort so frequently felt after eating
a large meal, when there was no demand in the system
for it. An overloaded stomach acts like a horse with a
heavy load on a bad road—very slow. There is a chance
that it won't get through; and if it does, it will be in a
bad condition. The most common result of overeating
is to throw a great bulk of gaseous, fermenting food
into the intestines, unduly distending them and preventing their action, which is a common cause of constipation.
Under such conditions, digestion will be very imperfect,
and the system poorly nourished and burdened or poisoned by the waste.

Bad Cooking.

Next to overeating, it is somewhat difficult to determine
whether bad cooking, folly in drinking, or haste in eating,
causes most disease. All are well-nigh universally practiced in this country. If our people had to do without
foods until they were properly cooked, most of them would
starve to death. Some modern cooks try to please the
taste, regardless of the stomach, while a large number
make no effort to do either. This class simply bring heat,
water, food and fat together, and trust to patent medicines and the doctor to keep alive those who eat their

products. The modern cook has been energetic in one direction, at least; and that is to get as far away from rational processes as possible. The object of cooking is to disintegrate the food and make it most palatable, but the cook often does everything possible to serve the food in such a way that neither end is attained. Heat coagulates or solidifies albumen—the principal element of meat and eggs—so, in order to make them as nearly insoluble as possible, the cook keeps them subject to a hot fire for a long time, and, as if this were not bad enough, they are often saturated with butter or lard. Fats are not digested in the stomach, and when meat or eggs are saturated with it, the food is sheathed with material not acted upon by the secretions of the stomach. Could any process be worse than to first render the food insoluble and then smuggle it through the stomach under a cover of fat? This is not all; for the process of cooking starch is equally bad. All starch-yielding foods are composed of layers of starch cells, from one-eight-hundredth, to one-five thousandth of an inch in diameter. These cells are enclosed in indigestible envelopes, which must be ruptured by cooking. Rolled oats is often served within five minutes after it commences to cook, while two hours' cooking would be nearer right. The method of cooking coarse, fibrous vegetables, such as cabbage, is almost as bad. The stomach is the disintegrating vessel of the body, and half-cooked, woody or stringy vegetables, saturated with fat, cannot be dissolved in weak stomachs. The ordinary cook spoils the meat by over-cooking; the starchy foods by almost no cooking; and caps the climax of cuisine folly by serving doughy bread. Verily, the cook slays not only thousands, tens of thousands, but millions.

Uselessness of Teeth.

The Creator either made a mistake in giving man teeth,

or man makes a mistake in not using them. The teeth were evidently intended to crush all solid substances before they were swallowed. It was also intended that saliva be mixed with the food to an extent of thorough saturation; but most people swallow their food without much crushing or saliva, and to make their folly complete, they wash it down with quantities of both hot and cold drinks."

Drinking Folly.

"If he could run like he can drink, I would like to hunt hares with him," can still be applied to many people. Folly in drink seems to have begun with human life, and we fear it will only end there. The use of alcoholic liquors has long been an important factor in producing disease. Strong liquors paralyze the nerves, deaden sensibility and irritate the mucous membranes of the stomach. Malt liquors, in quantities, derange the stomach, because of their icy coldness, their bulk and the amount of acetic acid they contain. Alcoholic inebriety is a great curse both physically and morally, but is not the only harmful drinking.

The reformers are not free from folly of a serious character; for many of them are tea and coffee inebriates, which, if not so bad as alcohol, they make up in the numbers they injure for what they lack in the intensity of effects. The habit of stimulating the nervous system with tea and coffee and tobacco, through its direct and hereditary influences, is one of the principal sources of alcoholic inebriety. Any considerable drinking during meals is pernicious, because it dilutes the digestive secretions too much, and makes too great a bulk in the stomach; but warm drinks are not quite as bad as very cold drinks, which lower the temperature of the stomach and paralyze its nerves. Ice water with meals is one of the most stu-

pid pieces of folly practiced by Americans (mostly men),
but a few (mostly female) have a still worse habit. They
drink liquids scalding hot. This practice leads to very
serious results, as the excessive heat irritates the mucous
membranes and relaxes the stomach and causes its dilatation.

Foul Mouths.

Some people keep their mouths about as foul as garbage boxes. The food adheres to the teeth and decays;
this destroys them, makes an offensive breath, and poisons the food to a greater or less extent, and causes it
to decay in the stomach.

Deficient Diet.

To maintain health our food must supply the chemical
elements found in the body, in proper proportions, and
as most people must select their foods from a very limited
number set before them, it often happens that the appetite causes them to eat those things that contain an excess of some elements, but deficient in others. Many eat
an excess of sugar, syrup, preserves, candy and sweetened foods. Many dyspeptics are cured by leaving off
all sweetened or sweet food. A large number eat too
much fat in the way of fat meat, butter, cream, gravy and
pastry rich from shortening. Free fats in the stomach
envelop the food and resist the action of the gastric juice,
and delay digestion. If fats remain in the stomach too
long, they are partly converted into butyric acid, which
irritates the stomach. A common instance of a badly-
chosen diet is excessive meat eating. If meat be eaten as
the principal part of the diet, it must be used for tissue-
repair and heat-production, or it will poison the body if
not promptly eliminated. Some go to the other extreme,
and eat very little but starch, white bread and potatoes.
Such a diet is deficient in protein and mineral matter.

Too Little Waste.

One of the most common faults or mistakes in our diet, is the use of foods that contain too little waste material. This comes from the practice of removing all the cellulose (bran) from our breadstuffs, so that they may appear white. Finely powdered bran acts as a stimulant of the bowels, and is the best of all known remedies for constipation. The practice of eating coarse bran, as advised by some physicians, is a great mistake. It is too irritating, and likely to remain in the stomach too long, and may also obstruct the bowels.

Incompatibility of Foods.

Foods may be good and wholesome enough, but incompatible when taken together. As an example, acids arrest the digestion of starch, and acids and milk will often cause vomiting. Strong tea makes eggs insoluble in the stomach, and both strong tea and coffee arrest digestion. Another common fault is that of eating too many kinds of food at the same meal. Some chemical elements unite; others will only mix like oil and water. Foods contain various chemical elements, and the fewer that are mixed at one time the better for digestion. Another thing, foods that are easily digested, and sour quickly, like sweet fruits and sweetened starch puddings, should not be eaten with hearty meals, or with foods that require a long period for digestion. If this be done, they are likely to sour the whole meal.

Eating too Little Food.

Many people, who are somewhat dyspeptic, eat too little food. Upon finding that many foods disagree with them, they restrict their diet until it barely sustains life. This increases their constipation, weakens the system and aggravates their dyspeptic troubles. (See Dietaries, for quantity required.)

Foods may irritate the stomach: (1) Because of their chemical composition. Such foods as raw onions and strong radishes are examples. Also such condiments as pepper, sage, curry, horse radish, mustard, and other pungent substances. But few people would care to blister the outside of their bodies; yet they have no hesitation about blistering the inside.

(2) Foods may irritate, because insoluble. Examples: Green fruits, raw, tough vegetables, pieces of nut kernels, lumps of meat or any other coarse, hard substance.

(3) Probably the most irritating of all foods, are those that contain poisonous ptomaines. The most common foods of this class are decaying fruits, decomposing milk products and poorly-prepared canned goods. Both fresh and canned fish are frequent sources of violent gastric attacks, because they are often tainted.

(4) Foods frequently contain mineral poisons from the vessels in which they have been kept.

Irritating Drugs.

The abuse of drugs is one of the most common and potent causes of disease; especially constipation and dyspepsia. It is putting it mildly to say that not a few have drug mania. Cathartics act by irritating the mucous membran of the digestive tract, and every time it is irritated, it is likely to become less sensitive, until finally the ordinary contents of the intestines do not stimulate it to action, which necessitates a constant repetition of irritating cathartics.

Irregular Eating.

Nature has method in everything, and we are naturally inclined to sleep and eat at regular periods. If we eat irregularly we break the rhythm of nature, and it is just as bad for the stomach to delay eating as it is for the quality of the dinner to delay after it is prepared. The

stomach cannot keep an adequate amount of digestive
fluids to be drawn on at any or all times. But this is not
all. Going an extra long period without meals causes
acute hunger and an overloading of the stomach; and an
extra short period does not give time for the stomach to
empty. Many people eat regularly during the week, but
on Sunday they disturb the regularity of the system by
eating a hearty breakfast from one to three hours later
than usual, and then an extra large dinner within four
or five hours. It is this pernicious practice that makes
people so uncomfortable and dull on Sunday and unre-
freshed to begin a week's work on Monday.

Evacuation of the Bowels.

A frequent cause of disease in the cities is neglect to
evacuate the bowels. Pressure of work and laziness is
assigned as a reason. The calls of Nature should always
be promptly met.

Loss of Sleep.

Sleep and rest are both necessary; without them the
nervous system has no tone and there is general languor.
Late hours burn the candle of life at both ends, and
night revelers sooner or later pay a fearful penalty, for
late-hour gaiety.

Excesses.

All excesses disturb the system. This is especially true
of those of a private character.

Tobacco.

The use of tobacco is both vile and pernicious; and the
physical wrecks, directly or indirectly due to tobacco,
would astonish the world if all were bunched together.
Dyspeptics should quit tobacco.

Dress.

A distinguished New York specialist reports that his
books show that thirty-four per cent. of the women treated

have displaced stomachs, while only six per cent. of the men were so afflicted. Of the females treated, the principal factor in the displacement was the corset. Of the males, probably the excessive use of beer and other liquids was responsible for their condition. Figure II shows side-view of female form. The dotted lines show the outline of a "neat waist" and "fine form," after the adominal organs have been displaced by the corset.

Fig III shows front view. The outside waist lines represent the natural waist; the middle dotted waist line the waist of the ordinary corset-wearer, who says that she does not "lace;" the inside dotted waist line shows how much the corset can improve Nature.

Figure IV shows natural position of abdominal organs.

Figure V is a front view of female, showing abdominal organs as displaced by corset-wearing.

Heredity.

No one can doubt the tendencies of heredity. Where both parents are weak their children will likely be so too; and if there be many children, some will be stronger and some weaker than their parents. The chief crime of parents against Nature is in transmitting nervous tendencies that make the child seek stimulants. How could it be otherwise, when parents use stimulants from the time they could talk? Can you blame the child of such parents who cries for strong coffee? By beginning early in life, hereditary weakness can, in the main, be overcome, so that personal responsibility can not be ignored, because of the transgression of our parents.

Local and General Diseases.

Local and general diseases are often most strongly manifested by disorders of the stomach. This is especially true of lung diseases—bronchitis and consumption. In these diseases, repugnance to fats is often one of the first symptoms.

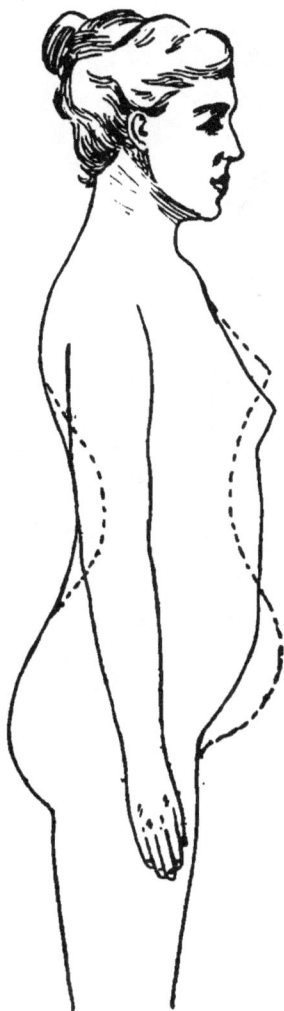

Fig. II.

Side view—female form, showing natural outlines.
Dotted lines represent the change due to lacing.

329

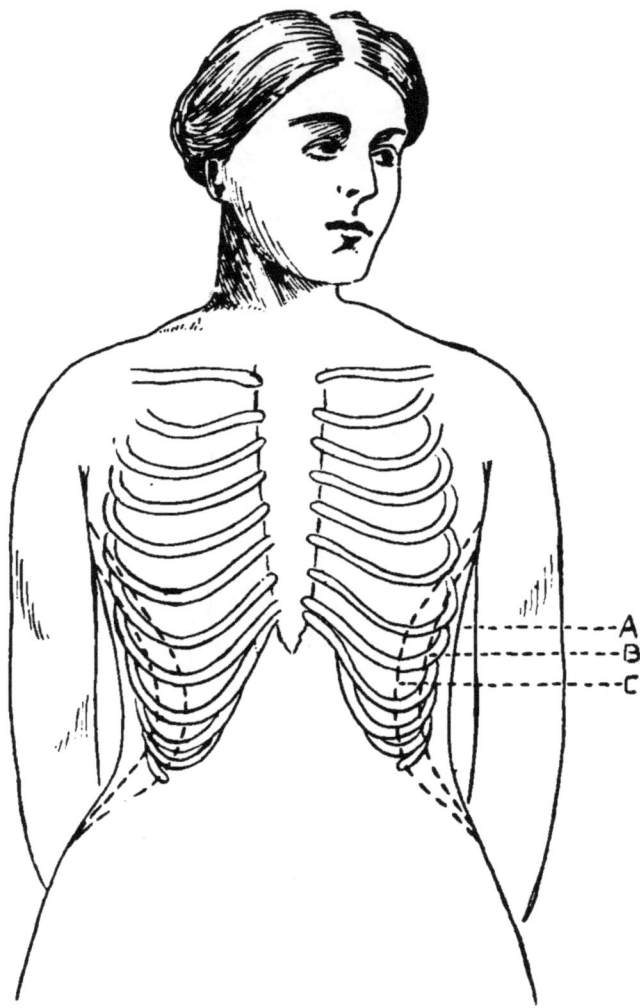

Fig. III.

Front view—female form.

A. Outline of natural waist.
B. Outline of ordinary form, "not laced."
C. Common outline of fashionable waist.

330

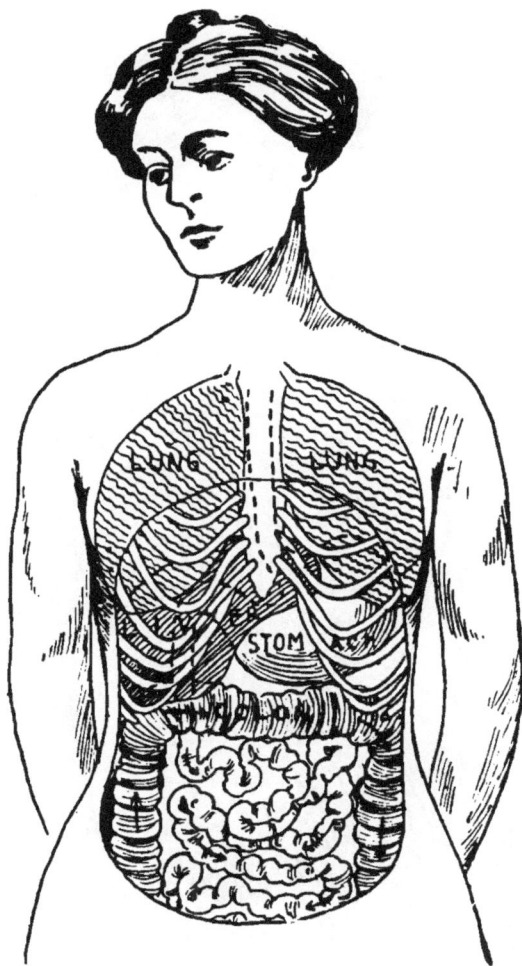

Fig. IV.

Front view—female form, showing natural position of abdominal organs, lungs, etc.

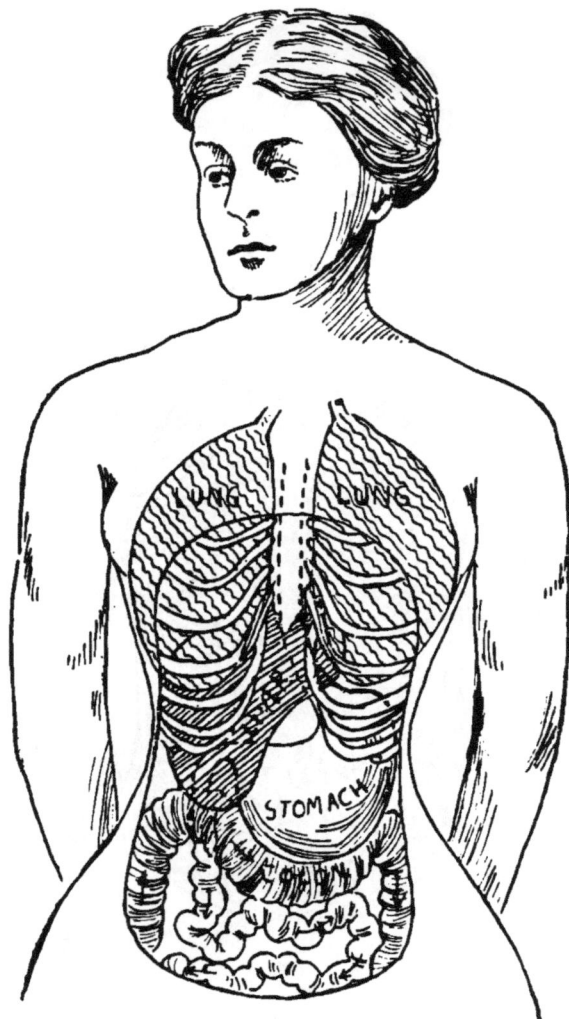

Fig. V.

Front view of female, showing displacement of abdominal and other organs, due to corset wearing.

Heart Disease.

Indigestion from diseases of the heart is doubtless due to disturbance in circulation; but it should be borne in mind that the heart is much more likely to be affected from the stomach, than the stomach is from the heart.

Diseases of the Liver.

The stomach is probably more in sympathy with the liver than any other organ; not so much from the assistance it gives in digestion, but because it is the chief organ for removing poison from the blood, which may cause indigestion in two ways: (1) From imperfect elimination of nitrogenous waste. (2). Reflexly, by constant irritation. Of the latter, painters, workers in lead, people living in newly painted houses, are the ones chiefly affected. Formerly, lead water pipes were sources of poisoning, but are not so as now made; but the same cannot be said of face powders. These arrest digestive secretions, and greatly diminish the churning movements of the stomach.

Diseases of the Intestines.

Diseases of the intestines are almost certain to affect the stomach, probably because excessive activity hurries the food out of the stomach before it is digested, and lack of activity has the opposite effect. It may still have another effect, resulting from imperfect digestion. Intestinal digestion being very important, if it fails, the blood will not be supplied with necessary elements, and the stomach may show the ill effect.

Malaria.

Malaria is likely to affect digestion from several causes. Fever of any kind diminishes digestive secretions of the stomach. In addition to this, if the liver is overworked, it cannot perform its functions perfectly.

Diseases of the Throat.

These may affect the stomach in two ways: by reflex irritation and from pus, or mucus, being carried to the stomach.

Anything that irritates the fauces may cause vomiting and tickling the throat has long been practiced for that purpose.

Pregnancy and Female Disorders.

Pregnancy is frequently accompanied by stomach disturbance; especially nausea, vomiting and craving for particular, and often peculiar, articles of diet. With female diseases, there is frequently associated some stomach trouble which results from the particular disease and disappears with it. Thus the physician, in treating women for stomach trouble, should ascertain whether or not it is simply stomach disease, or secondary to some other female disorder.

Heat and Cold.

Excessive heat and overexertion cause general collapse; but the indigestion, common to hot weather, is mainly due to cold drinks and use of decaying fruits. Cold, chills the surface and disturbs the circulation, causes congestion, and partially arrests the elimination of waste.

Pressure on the Stomach.

This is common in occupations requiring a stooping attitude. It restrains the natural activity of the stomach and interferes with digestion.

Light.

The importance of light is often overlooked. Man was never inteded to live in a cave or dungeon and work by artificial light, and those who do so pay a severe penalty. Ordinarily, people are not seriously affected by slight violations of the principles of good living, if not too long continued, nor too many of them. In most cases of illness it will be found that several causes operate together to produce the disease.

CHAPTER XXVIII.

DISEASES OF THE STOMACH.

Diseases of the stomach are classified as follows:

Nervous dyspepsia, neurosis of the stomach, gastric neurasthenia and gastralgia, are names applied to various affections of the stomach that have their origin in the nervous system. Acute and chronic gastritis designate acute and chronic inflammations of the stomach of a catarrhal character. Hyperchlorhydria and hypersecretion, apply to excessive secretion of hydrochloric acid in the stomach. The former is used to designate the excessive secretion of acid during meals; the latter to uninterrupted secretion without any relation to meals. Ulcer is a sore on the lining membrane of the stomach, which, in some cases, may perforate it. A dilated stomach is one that is stretched beyond its natural size and remains so. Cancer is a tumor of the stomach, which grows more or less rapidly, and interferes with digestion. This classification is made according to the manner in which the stomach is locally affected, rather than the cause of the disease or the symptoms produced. Neither of the diseases named have all symptoms entirely different from other diseases of the stomach, but usually each has some distinguishing characteristic. The time is past when physicians can call any disease of the stomach dyspepsia or indigestion, and stop there. The modern doctor must now determine what kind of indigestion his patient nas, and to do this it may be necessary to take out the contents of the stomach and make a chemical examination of it. The fact that many physicians have not been able to differentiate one disease of the stomach from another, ex-

plains why so many dyspeptics have failed to be benefited
by medical treatment.

Nervous Dyspepsia.

This ailment is not really a disease, only a local manifestation of some nerve derangement. It differs from
all other diseases of the stomach in this: it has no
anatomical change and is not directly due to any alteration in size or structure of the stomach, but to some shock,
strain, nervous exhaustion, or nerve irritation in organs
other than the stomach. Some eminent physicians are
disposed to charge all but contagious or infectious diseases to some functional disturbance of the nervous system, but it seems that in most cases the theory of nervous
origin is, "Putting the cart before the horse"—the effect before the cause. Strictly speaking, the term nervous dyspepsia should only be applied to diseases or symptoms which
result from mental disturbances, such as emotional excitement, mental worry, or mental activity, too long continued without rest; but for practical purposes it is better to class all derangements of the stomach directly due
to the causes mentioned, or that are simply reflex from
nerve irritation in other parts of the body, as neuroses of
the stomach. Some writers have heretofore classed both
insufficient and excessive secretion as of nervous origin,
doubtless because they were attended by nervous symptoms. All agree that it is extremely difficult to determine
the dividing line between cause and effect. We have in
the preceding pages shown how the system may poison
itself, and if the blood contains crude or poisonous matter,
it is very likely to affect the nervous system. If the food
supply be of such a character that it cannot be digested,
the whole system will be weakened, and fatigue follow
very little exertion, and if lack of nourishment affects the
muscles and the sensibilities, why not the central nervous

system? On the other hand, a debilitated nervous system, from extraordinary worry or loss of sleep, affects the stomach. As to whether the nervous system or the stomach is the primary cause of the nervous symptoms, would seem to depend on the causes in operation likely to produce them. If there were some cause of extraordinary worry and loss of sleep, the indigestion should be considered as of nervous origin, but if there were a disposition toward unusual worry about trifles in connection with nervous symptoms, it is most likely due to some form of mal-nutrition, or self-poisoning. But few people would be overworked or over-worried if they were properly nourished; and by this we do not mean that there has necessarily been a lack of nutritious foods. It may be because of the inability of the system to assimilate the food as it is supplied. As people mistreat their stomachs much more than their nervous systems, it is safe to consider that, primarily, it is not overwork that causes most nervous attacks, but lack of good blood and excess of poisonous waste in the system. This is well illustrated by the general anaemic condition, common to most persons suffering from nervous dyspepsia. The disease is much more common to women than men, and more frequent under forty years of age than after the middle period of life.

Nervous dyspepsia has no uniform symptoms, as there may be lack of gastric secretion and muscular activity, an excess of secretion, or extreme irritability of the nerves of the stomach.

Symptoms.

Where there is lack of secretion, or lack of muscular activity, the symptoms are much less marked than the other conditions. Vomiting only takes place when the food has long remained in the stomach and become de-

composed.   In the mild form there is not much pain, but
a sinking sensation, or one of great fullness is felt after
eating, which may be accompanied by slight nausea and
dizziness.   In severe cases the stomach will not tolerate
any food at all, and vomiting occurs almost as soon as the
food reaches the stomach.   Some of the symptoms of
nervous dyspepsia are found in other diseases of the stomach, but there is usually something about each that indicates to which class it belongs.   In catarrh of the stomach
the disagreeable sensations do not arise until some time
after the meals, usually three or four hours, unless there
is an acute attack and great irritation of the stomach.   In
some cases there is excessive sensitiveness of the mucous
membrane of the stomach, and the pain severe and the
stomach sensitive to pressure.   The pain is more general
than in ulcer, and has but little relation to meals or kind
of food, while the pain from ulcer is much more intense
when coarse vegetables or acids are ingested, than when
the stomach is empty or when soft-proteid foods like milk
and eggs are eaten.   In the form of nervous dyspepsia
known as gastralgia, the pain in the region of the stomach seemingly radiates in all directions.   It occurs quite
independently of meals.   In neurasthenia, the symptoms
and pain are generally out of proportion to any discoverable disease, and often occur when there has been no
previous history of dyspepsia.   Another characteristic of
nervous dyspepsia is belching of air or gas, without regard to whether the stomach is full or empty.   In other
diseases of the stomach the belching only appears when
there is gaseous fermentation of food in the stomach.
When more or less food is brought up instead of gas, it
is called regurgitation, which often precedes nervous vomiting.   Sometimes the openings of the stomach are closed
by a nervous spasm, or the pyloric end may, for a time,

refuse to close and the food at once passes into the intestines, causing diarrhœa.

Vomiting in Pregnancy.

This is usually called "morning sickness," and appears after rising in the morning, when the patient feels faint, "light-headed" and nauseated. When this occurs, slowly sip a cup of hot milk or meat broth, and eat a dry biscuit (cracker) and remain in bed for two or three hours. If the stomach be foul, drink a cup of hot water instead of milk, without food.

Aids to Treatment of Nervous Dyspepsia.

If the attack be due to overwork or worry, rest is the first requisite; but if from emotional excitement, change in surroundings and something to divert the mind is of great importance. When there is general neurasthenia or hysteria, the patient should be put to bed and kept free from excitement and away from visitors.

Diet.

The diet must be easily digested, nutritious, and non-irritating. If the stomach be inactive, so that it does not readily empty itself, the diet must be of such character as will not quickly ferment. The principal foods should be malted milk—pasteurized or sterilized—cream, soft-cooked or whipped eggs, eggnog, malted gluten, meat broths, in acute cases. Such additional foods as meat-powder, toast bread, nut oils, butter, malted nuts, may be given as the patient progresses. If hot milk should be vomited, try it cold, and vice versa. It may be of great advantage to dilute milk with gelatine. If not convenient to make it, a refined gelatine, like the Keystone, may be used. As the appetite is often capricious, it may be of great advantage to flavor the food with a little vanilla, lemon, nutmeg, or fruit flavors. In the acute or severe cases, a little food should be given at short periods. It may be necessary to begin with a tablespoonful of milk. The patient must take as much nourishment as possible, but must not be crowded beyond what can be digested. When only milk is fed, it should ordinarily be given every two hours, in quantities of one or two ounces at a feeding for the first few feedings. Most patients will tolerate a pint and a half of milk the first day, and twice as much the second, and should be able to take eggs and other food the third or fourth day. All made dishes, tea, coffee, and fried foods, must be avoided.

Acute and Chronic Gastritis, Usually Called Catarrh of the Stomach, or Bilious Attacks.

These are the most common of all diseases in the United States, except colds. Very few people escape occasional gastric attacks, although they may not be willing to admit the fact. It is indeed a strange thing that people will insist that almost anything ails them except some form of indigestion or mal-nutrition.

Gastritis, or catarrh of the stomach, is the every-day dyspepsia of the world. Its causes, briefly re-stated, are as follows:

Excess of food, incompatible, irritating or decomposing food, poison taken in or originating in the body, excessive heat, and disease of other organs, especially the intestines.

Symptoms.

In the ordinary catarrh of the stomach any of the following symptoms may be felt:

Headache, offensive breath, "bad taste" in the mouth, drowsiness, nausea, loss of appetite, great thirst, vertigo, vomiting, belching some hours after meals, constipation or diarrhoea, lassitude, aching limbs, cramps of the muscles of the leg, pain after eating, flatulence, heartburn, difficult breathing, palpitation of the heart, stomach feels like it had a weight in it, tenderness of the stomach, eruption on the lips, tongue raw, red or coated, lack of energy, chilly sensations and coldness or numbness of hands and feet.

The acute attacks are generally called "bilious attacks," and occur most frequently in the night. The patient will usually be wakened by pain, and in some cases there is a feeling of nausea, followed by vomiting and relief. They may occur from only slight or accidental causes, but when the ailment becomes more or less continuous they are then termed chronic. Gastric attacks are often so severe that the patient thinks death imminent, although in no danger whatever. Acute gastritis occurs at all ages, while chronic gastritis usually occurs in middle age or late in life, due to slowly-progressive indigestion. In acute cases there is only congestion; but when chronic, there are structural changes in the stomach, deficiency in digestive secretions, an excessive secretion of mucus, loss of absorptive power and muscular activity. When this condition exists, foods difficult to dissolve, like salt lean meat, especially when fried, and coarse vegetables, will disagree. This will also be true of foods that ferment quickly, such as custard puddings, tapioca, sweet and sweetened fruits, vinegar with starch of any kind. Milk, without any alteration, will usually disagree, because the stomach is not sufficiently active to break up the curds.

Diet.

In extremely severe cases it may be necessary for the patient to live on milk, diluted with gelatine, or barley water, or it may be malted and used to great advantage with malted gluten. All fried foods must be eschewed, and except in acute attacks, all soups, mushes and gruels must be sparingly used. The diet must consist mainly of dry foods, thoroughly masticated. Saliva and thorough mastication will do more for a damaged stomach than almost all other remedies. The stomach must be strengthened, by giving it as much work as it can do and no more. It can never get strong without plenty of nourishing food, and "slops" will not do. The curse of the American stomach is slops, water and chunks. This unfavorable mention of water would probably please a Kentucky colonel, but the objection is only in the manner of using it. Dyspeptics need to eat as great variety of food as possible, but it must be done discreetly. Foods that ferment quickly, must be avoided, or when used, it must be on an empty stomach, or with easily-digested foods. If the digestive secretions are deficient, meat will be poorly tolerated, and an exclusive diet of meat and eggs for two or three days will determine this. The cereals are the best reliance, although if complicated with severe intestinal disorders, only gluten should be used, with such foods as meat, milk and eggs. As an aid to curing constipation, there is nothing equal to the bran of cereals, when finely ground. If not convenient to take foods containing fine bran, it will be advisable to boil it for three hours, then roast until brown. It should then be ground as fine as possible. If desired, flavoring matter may be added to make it palatable. Graham bread is objectionable, because the bran is too coarse. Where the stomach is greatly inflamed, sour fruits are not allowable, but in chronic cases of mere sluggishness they are of great benefit, if eaten at proper time, without sugar. ft too sour, a little bicarbonate of soda (baking soda) may be added while cooking. Sweet fruits may be eaten when the stomach is empty, but if it contains the residue of a meal that has soured, they will quickly produce flatulence. When the stomach is very weak, it will be necessary to take small quantities of food every two or three hours during the day; but in chronic cases, where the stomach will do its work, if given plenty of time, two meals a day, eight or nine hours apart, will be far better than three or more. The patient must early learn that a suitable diet will do far more to effect a cure than any drugs. As an artificial aid to digestion, very good results are sometimes obtained from malt tea. It may be made as follows:

Take three or four large tablespoonfuls of malt and steep it in a half-pint of cold water ten or twelve hours. Decant, bottle, and keep in cold place. One or two tablespoonfuls may be used at meals with a little milk and hot water.

Diseases of Excessive Secretion. Hyperchlorhydria, Hypersecretion and Ulcer.

These diseases are closely related and usually represent the first, second and third stages, although it is claimed that hypersecretion may commence suddenly.

Causes of Excessive Secretion.

Excessive use of alcohol, mustard, pepper and other condiments. The use of ices. Too rapid eating for a number of years. Indigestible food, grief, worry and pressure on the stomach. The disease may commence suddenly or gradually.

Symptoms.

The principal symptoms in excessive secretion is pain. It may begin with mere uneasiness, one or two hours after meals, or with sharp, stinging pain. The excessive amount of acid irritates the stomach, and as soon as digestion in the stomach is completed (usually from two to three hours) the pain begins. In severe cases there may be an attack after every meal, the one after breakfast will be the lightest, and the one after dinner the most severe. Often a little hot acid liquid will be belched. The pain is often sharp and severe, and is usually called cramps. It may be relieved by taking a drink of water, which dilutes the irritating acid, or by eating a soft egg.

Appetite is usually good and the tongue clean. There is no flatulence; no feeling of fullness. If meat or eggs be given every three hours, and it is well tolerated, it is suggestive of excessive secretion, because meats are slowly digested, when the secretion is deficient. If nothing but starch be taken when there is excessive secretion, it may remain in the stomach a day or two. The symptoms of excessive secretion are different from other diseases in this: In nervous dyspepsia, there is no time relation to meals. In gastritis there is more nausea and flatulence, a furred tongue, and the pain less sharp, and not relieved by food.

This is, ordinarily, only an advanced state of excessive secretion, although some specialists say that it may begin suddenly, but, in such cases, it is more than probable that it is a sudden manifestation of what has long existed. The symptoms are similar to excessive secretion, only more pronounced. Hunger is more acute. Patient may wake up in the night with an "all gone" sensation, and if nothing be eaten, there will be severe pain. Thirst is constant, especially at night. The attacks are generally worse in the middle of the night, and last two or three hours, and terminate by vomiting, which relieves the pain. Another characteristic symptom is diarrhoea in the night, due to the excessive acid condition of the food discharged into the intestines, and to the large quantities of fluid drank. This may be followed by constipation. Notwithstanding the voracious appetite, and large amount of food eaten, the patient usually gets thinner. The tongue is seldom furred, and likely to be very red. Probably the easiest way to distinguish hypersecretion from other diseases, is by the matter vomited. If the vomit shows that the lean meat is practically all dissolved, and the bread-stuff unchanged, it points strongly to hypersecretion. Where the excessive secretion has long been continued, the stomach is almost certain to be dilated.

Diet.

A meat diet is usually prescribed in excessive secretion, on the theory that meats are easily digested in an acid stomach, and starches difficult. This is true. And it is also true that meat furnishes the system more hydro-chloric acid than any other food. Now, in excessive secretion, the object is to reduce it, and what more rational method can be proposed than to withhold foods that make most hydrochloric acid? Diet in this disease must be as bland as possible, and as milk is well tolerated, if diluted as heretofore described, it is the best of all foods. Of meats, fresh fish is the easiest digested, and of most service in excessive secretion.

In catarrh of the stomach it is necessary to have the food finely divided, so that it can be dissolved. In excessive secretion, it is necessary to have it as fine as can be powdered, so that it will not irritate the stomach, and excite the secretion of more acid.

The chief difficulty is in the digestion of starches, and it is not easy to prevent loss of weight without them. Continued loss of weight, means loss of strength, and great care must be taken to maintain it. Bromose (malted nuts), cream, nut oils, and the fat of boiled ham, will be useful in furnishing fuel for the body. If the stomach can be washed out once a day, considerable dry toast may be eaten soon after, if taken without any liquid, and thoroughly mixed with saliva. Malt will also be serviceable, and malted gluten should be used in preference to meat or meat powder. Sour fruits are not suitable, but such fruits as bananas, sweet grapes and pears can be eaten, unless there be dilatation of the stomach. In such cases foods that ferment quickly must be avoided. The cereals must be extra well cooked, and then roasted brown and powdered, and then eaten dry. If the stomach is not washed, a half pint or pint of cold water or moderately cold alkaline mineral water may be drank a half hour before meals and before retiring at night. Mucilaginous drinks, made by steeping "slippery elm" (ulmus fulva) in cold water, may be drank before or after meals.

If there is no dilatation of the stomach, the patient should eat as often as every four hours. The bowels should be kept active by massage. All irritating substances such as pepper, mustard, raw vegetables, vinegar, sage and cheese, must be shunned as enemies. A few grains of salt may be used in the food, but the less the better. Very hot drinks are absolutely prohibited, especially where there is a possibility of ulcer.

Ulcer.

The question is sometimes asked, why don't the stomach digest itself? If an animal or human being be suddenly killed during digestion, the stomach will digest itself to a considerable extent. In the living stomach it is supposed that a continuous supply of fresh blood protects it from its own secretion. It would follow that if the circulation in some part was partially, or wholly destroyed, that the stomach might dissolve itself. In excessive secretion, the strong acid probably erodes the membranes in the stomach, which develop into ulcer, if the causes producing the erosion are long continued.

Probably the principal cause is pressure on the stomach, from a faulty system of dress. Other causes are indigestion, irritating foods, and a general neurotic condition of the system which seems to be closely associated with ulcer. It is most commonly found in females between the ages of twenty and forty.

The common location of ulcer is near the pyloric end of the stomach or on the anterior wall. It is also found in the intestines. The most characteristic symptom is pain at the exact spot of the ulcer, and immediately opposite in the back. The stomach is often sensitive to pressure from other diseases, but in ulcer it is particularly sensitive to pressure at the exact spot where it is located. The pain follows the indigestion of foods, and bread or vegetables give much more than milk or soft eggs. If vomiting takes place, the pain is relieved. One of the symptoms most relied on is hemorrhages, and when blood is vomited, and attended by the other symptoms common to ulcer, it is almost conclusive that ulcer exists. The patient generally grows progressively thinner, unless properly treated, and when fatal dies of starvation or perforation. There is a form of ulcer called peptic ulcer, in which the characteristic symptoms of ulcer are absent. These cases are rare, but extremely difficult to diagnose.

A patient with ulcer of the stomach should be put to bed, and no foood given in the regular way, except sips of ice water or cracked ice. The nourishment must be administered through the rectum, until the ulcer heals. The rectum must first be cleansed by an injection of water, and then about three to four ounces of pancreatinized meat powder, milk, or milk and eggs pancreatinized, should be administered. Some use a 20% solution of sugar, beaten with three eggs. Whichever nourished the patient most should be used, but ordinarily eggs and milk, equal parts, with a pinch of salt, will be found most useful. Five or six feedings a day will be necessary, and one of them must be water, to be retained for absorption, as water is as necessary as food.

When the stomach is healed sufficiently, feeding may be resumed by giving a teaspoonful of cold milk to begin with, but it would be advisable to first cleanse the mouth with some antiseptic wash. A few grains of powdered boracic acid with a little water and tooth brush will answer. If a teaspoonful of milk is tolerated, two teaspoonfuls may be given at the second feeding and so on, increasing slowly, until the patient can take four or more ounces at a feeding, every two or three hours. As feeding by the stomach is resumed, the rectal feeding should be discontinued at a rate of one feeding a day. The principal diet in ulcer is milk and ice cream, which should be plainly made. It would be well not to be too hasty in increasing the diet in ulcer, for it may be necessary to live on milk and ice cream for some months. The first addition to the milk except milk diluents, that is allowable, is one egg beaten and eaten with the milk. Hot dishes are positively forbidden, as they are likely to cause hemorrhage. During convalescence the diet should be similar to that in excessive secretion.

Cancer of the stomach is so uncommon, as to scarcely deserve mention. Various theories as to its origin have been proposed, but they are purely speculative. The origin is unknown, further than that it appears to be a hereditary tendency in a few families. It usually appears in middle life or old age. In the early stages of cancer it is difficult to diagnose as the symptoms resemble other forms of dyspepsia.

Cancer has one characteristic different from all other diseases of the stomach. It is steadily progressive, and the end soon reached. There is a gradual loss of weight, tumor in the region of the stomach, frequent vomiting, and there may be either constipation or diarrhoea. Nearly all the diseases of the stomach are more or less intermittent, except cancer, and it is usually fatal in less than a year and a half. If the disease has been recurrent, or continued for a much longer time than stated, it is not a cancer. The diet should consist of milk, eggs, meat powder, nut meal, baked apples, cream and nut oils.

The stomach is an elastic pouch, and when it is filled with large quantities of food, water or gas, it becomes distended. This weakens its walls, and whenever it will not contract to its natural size, after being distended, it is said to be dilated. Prior to 1885, dilatation of the stomach had received but little or no attention, and the relation it bore to many diseases was unknown.

The direct cause is continued over-distention, from either food, water or gas. The injury of over-eating is well known, but over-drinking is equally bad, though not so common a cause of dilatation of the stomach. Distention from gas results from the putrefaction of foods due to indigestion or partial closing of the pylorus, called stricture. Whatever prevents the stomach emptying between meals, will cause abnormal fermentation and distention from gas. These causes have been discussed so thoroughly under causes of disease of the stomach, they need not again be repeated.

The symptoms are both local and general. It often happens that the stomach symptoms are not very pronounced, and both patient and physician are misled. Of the local symptoms, flatulency is the most common. When the stomach is most distended, the abdominal outline is greatly enlarged. If this be due to gases, the enlargement begins at the lower end of the sternum, but when due to the use of large quantities of water and beer the enlargement of the abdomen is lower. The different aspects compared with the natural outline (side view of male form) are illustrated in figure VI. Dotted line at A shows prominence of dilated stomach, beginning at end of the sternum, while dotted line at B merely shows dis-

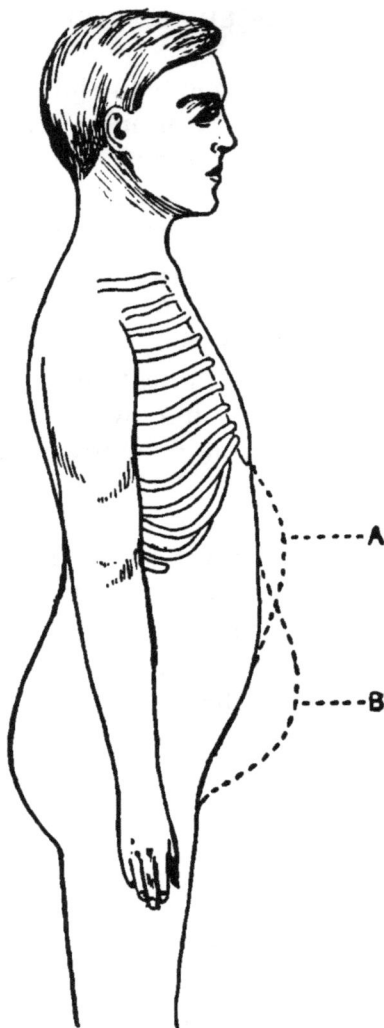

Fig. VI

Side view of male figure.

A shows prominence due to dilated stomach, without abdominal distention
B shows abdominal distention common in obesity, etc. Both conditions frequently exist in same person.

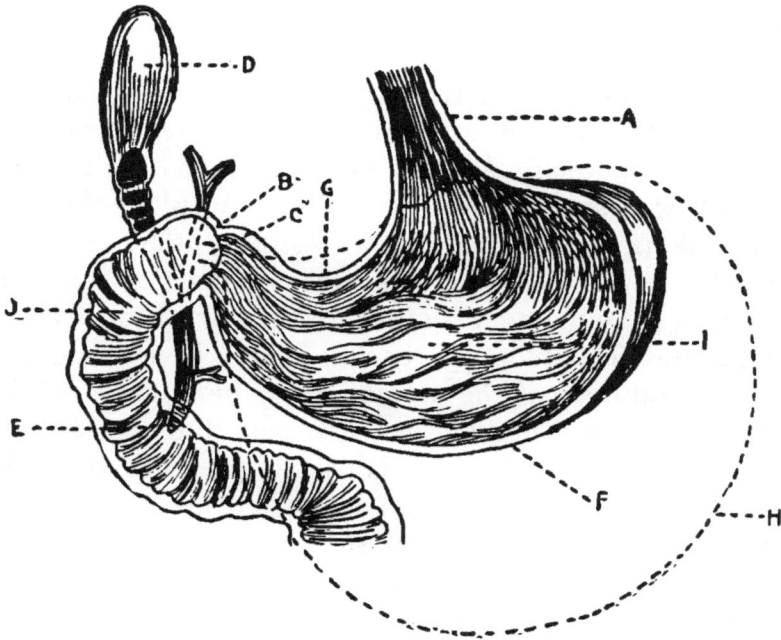

Fig. VII.

A Cardiac opening of the stomach.   B Normal pyloric end.   C Constricted pyloric end.   D Gall bladder.
E Opening of gall bladder into intestines.   F Greater
curvature of the stomach.   G Lesser curvature.   H Outline of dilated stomach. I Folds of stomach J douodenum.

tended abdomen. Where both conditions exist, the enlargement begins at A, and extends with increasing prominence to abdomen. We have seen a few cases where the stomach was so greatly dilated, that it would hold over four gallons. A simple way to estimate the size of a patient's stomach is to administer seidlitz powders unmixed, or by inflating with inserted stomach tube. Have the patient lie flat on the back. This will indicate the marginal outlines of the stomach. In most cases of dilatation, there will be but very little pain, but where there is stricture of the pylorus, it is usually preceded by ulcer and excessive secretion, which are both painful.

Figure VII shows natural stomach, and dotted lines indicate a dilated stomach due to stricture of the pylorus. In these cases the outlet of the stomach is narrowed by ulcers or inflammations, until the stomach is unable to empty. These can only be helped by surgical operation. In ordinary dilatation there is but little pain, but a feeling of fullness and weight at the stomach is almost constantly present. There is also frequent belching of gas, which begins two or three hours after meals. Sour liquid or food will often be brought up with the gas and "heartburn" will likely be a common symptom. The general symptoms of dilatation can hardly be enumerated. Appetite may be very good, excessive or poor. Tongue usually coated; person may be thin or corpulent; often the latter. Constipation is persistent, but sometimes alternated by diarrhoea. The feces are fetid, because of the putrefaction it has undergone. Dull head-ache and nervousness are common, and frequently there is great sensibility to cold. Exertion quickly exhausts. Persistent insomnia is a strong indication of a dilated stomach, and vertigo, unusual vision, dropsy of the limbs, numbness, cold extremities, flushing of the face, night sweats, asthma, neuralgia, eczema, are due to poisonous matter taken into the blood from putrefying food. One of the most alarming and sometimes fatal effects due to dilatation is palpitation of the heart. The enlarged stomach presses the diaphragm upward against the heart, causing heart failure. Many cases of sudden death in the night are accompanied with the announcement that the person ate a hearty supper, which probably caused the stomach to be distended with gas, and that in turn displaced the heart and caused heart failure. Too many people imagine that because their digestive organs do not double them over with pain, that they are all right, whereas if their food supply and digestion were all right, and the waste eliminated there would ordinarily be no distressing symptoms of any kind, except from contagious diseases.

Diet.

The diet must be free from bacteria, and of a character that does not quickly ferment. As a general rule, the digestive secretions will be deficient in a dilated stomach. This is especially true of chronic cases of long standing. Where the pylorus is partly contracted, so that the outlet of the stomach is reduced, there will likely be excessive secretion, and then the stomach has to contend with its own acid, and those due to excessive fermentation. These cases however, are not the ordinary ones. When the stomach is dilated, plain milk will usually disagree. It should be taken with a teaspoon, and each spoonful kept in the mouth a while (a minute or two). The food should be dry, and no drinks taken with it. All starchy food should be cooked an extra long time, and cereals should be both boiled and roasted. Fresh doughy bread must be avoided entirely, and all bread should be thoroughly baked, and should then be sliced, and baked again. Sugar and sweet fruits, sweetened pastry, syrup, preserves, jellies, and all sweetened foods, must be kept out of the dietary. Cooked, roasted, ground and crushed cereals, should be substituted as far as possible for bread, so as to avoid the yeast ferment and baking powders. Granose biscuits are as good a food for dilated stomachs as can be found. All cured meats and preserved foods are to be avoided, also all fried foods. Meat and eggs can only be partaken of in small quantities, and must never be fried. Peas and beans will be too solid, unless ground. So will nuts, but finely powdered nut meal with sour fruits will often agree better than any other food. Such stimulating foods as cooked onions without fat may be useful. Finely ground wheat bran will be a great aid, as it stimulates the digestive organs without obstructing or causing an excessive irritation. All stale fruits or other foods, except bread, are likely to ferment quickly, and are therefore not suitable. Rancid butter, or hot butter is bad under any circumstances, and particularly so for slow stomachs. Sterilized cream, nut oils, and the fat of ham or bacon, should be the only fats used. All free fats are objectionable, and this excludes all gravies. Malted gluten can always be used, and malt tea is an aid to starch digestion. The patient must be encouraged to eat plenty of food, as too little food means loss of strength, to resist disease. This can be accomplished by variety of foods and use of flavors. It would seem to be hardly necessary to say that, pickles, pastry, condiments, tea, coffee, tobacco and beer should be left for those who have no regard for their own welfare. In the beginning, the dietetic treatment of dilatation it is essential to first cleanse the stomach. This should be done with a siphon, but if it is not done, the next best process is to eat a light dinner, a light supper, and then, on the following morning, drink a half pint or more of alkaline mineral water, an hour before breakfast, and then knead the stomach and abdomen for at least twenty minutes. If mineral water is not obtainable, a suitable substitute should be prescribed. No headway can be made as long as there is foul matter in the stomach. Hot water drinking will usually do more harm than good, because the stomach is already too much relaxed. After the stomach is cleansed, the mouth should receive attention. The food must not be contaminated by decaying matter in the mouth. Lemon juice will cleanse the tongue and membranes, but it should not be swallowed. For the teeth, any suitable wash may be employed or soap, and a brush will answer. The teeth must be kept in condition to masticate the food, and they must be well used for that purpose. After the stomach and mouth are thoroughly cleansed, begin the diet with sterilized or pasteurized milk and malted gluten; or a soft boiled egg with granose biscuit or some other whole wheat food, that has been twice cooked. Meals must be regular and not closer than eight hours apart, except in acute attacks. Exercise, baths and pleasant surroundings are all important aids, and should be combined with regular habits. Every patient must be impressed with the fact that the diet and habits are far more to be relied on than drugs. The notion that chronic stomach troubles can be cured by drugs alone cannot be too quickly dispelled.

CHAPTER XXIX.

DISEASES OF THE INTESTINES, PANCREAS AND LIVER.

We have heretofore described (Chapter II) the general structure and functions of the intestines, but we might again emphasize the fact that a very important part of digestion takes place here, and that the intestines are not a smooth channel for the escape of waste, but have numerous folds and tongue-like projections which are so constructed that they absorb portions of the food which is then mixed with the blood, and carried mainly to the liver.

Now if the food remains in the intestines too long a time, poisonous matter may be generated and carried into the general circulation. This may also result from imperfect digestion, as the entire intestinal tract contains bacteria of various kinds, some of which are likely to be very active when there is any defect in natural processes. The ordinary diseases of the intestines are such as hinder digestion, absorption and the elimination of waste. Chief among these is constipation, and it can be truly said that there is no ailment in America so common.

No definition would seem to be needed for a disease so nearly universal, and yet there are many mistaken notions as to what symptoms clearly indicate its existence. It is generally supposed that daily, or at least frequent stools, is conclusive, that no constipation exists, whereas nature may make frequent attempts, yet be unable to fully clear herself. Whenever the waste of the food is not regularly and completely discharged, there is constipation, no matter how frequent the stools.

These may be enumerated as follows:

1st.   The habit of eating food that contains too little waste, ordinarily called cellulose.

2nd.   Imperfect mastication of food—too rapid eating.

3rd.   Ice water, and iced drinks generally, also iced foods.

4th.   Failure to drink sufficient fluids between meals.

5th.   Improper admixture of foods.

6th.   Over eating, i. e., eating so much food that the stomach is unduly distended.

7th.   Eating too little food.

8th.   Astringent foods and drinks.

9th.   Insufficient exercise.   ,

10th.   Lack of peristaltic movements of the bowels. -

11th.   Neglect to evacute the bowels daily.

12th.   Irregularity in eating.

13th.   Hereditary weakness.

14th.   Insufficient secretions of the various digestive organs.

15th.   Eating coarse, insoluble substances.

16th.   Over-distention of the abdominal wall.

17th.   Malaria.

18th.   Drugs, especially cathartics, opiates and astringents.

19th.   Mental influences.

20th.   Chronic diseases of the mucous membranes.

21st.   Excessive perspiration.

22nd.   Dilatation of the rectum from repeated injections of large quantities of water.

23rd.   (a) Pressure on intestines.   This may come from tight waist bands, corsets, belts or clothing.

(b)   Stooping posture, common to seamstresses, students, bicycle riders and others.

(c)   From pregnancy.   This results both from pressure and from general inactivity of the bowels, common to this condition.   ,

24th. Disease.

The effects of constipation are so far-reaching, as to be well-nigh indescribable—they cannot be enumerated. There is always danger that toxic substances will be absorbed into the system, which may cause merely a feeling of discomfort, or disease of any organ of the body. Aside from this, accumulation of fecal matter will likely dilate the intestines so that they permanently lose their contractile power. This is one of the reasons why chronic constipation is so hard to cure. It is also a source of pelvic and genital irritation out of which arise immorality and even crime. Hemorrhoids (piles) is another common effect of constipation, although they may have other causes. The most common symptom of constipation is general lassitude. When toxic substances, caused by retained fecal matter or mal-fermentation of foods, are absorbed, the nervous system is quickly affected. Sometimes there will be increased activity, and a feeling of unusual vigor, followed by flushes of heat, which cause the face to burn. This, the laity often mistake for fever, and the general symptoms of auto-intoxication are often supposed to be malaria even by physicians in high standing. Other manifestations of constipation and self poisoning are headache, indolence, dullness, sleeplessness, stupor, loss of appetite, vertigo, burning sensations in the stomach, tenderness of the gastric region, foul breath, flatulence, palpitation of the heart, pain in the back, and a moody and irritable disposition.

The use of cathartics—except immediate results be absolutely necessary—must be discontinued. This also applies to all habits that are contrary to proper living as heretofore laid down. Meals must be regular every day, including Sunday, and ordinarily should not be closer than six hours for three meals a day, and eight hours from breakfast to dinner, with two meals a day. This is imperative. Foods must be properly cooked, and incompatible foods, though wholesome, must not be eaten at the same time. The practice of eating sour fruits and then oat meal and milk, is a good illustration. The porridge is likely to get but little saliva, and its action be immediately arrested by the acid previously taken. Salivary digestion will ordinarily be arrested in the stomach, by its secretions, in twenty or thirty minutes after food is ingested, and no acid should ever be taken with starchy foods, nor sooner than twenty minutes afterward.

It is very important that all food be thoroughly masticated, because it facilitates digestion, prevents gaseous distention and obstruction of the bowels. The quantity must also be adapted—neither too little nor too great excess. Too little food weakens, and too much over-works and in a sense, paralyzes. No food exerts as good effect on sluggish bowels as fine bran. Coarse bran used in graham bread—frequently prescribed by physicians—sometimes obstructs the bowels and leads to grave consequences.

Next to over-eating, the habit of drinking large quantities of coffee, tea, or ice water, with meals must be avoided. The digestive secretions must not be greatly diluted. Small quantities of moderately warm fluid may be taken with meals, but the habit of pouring in liquids three times as fast as they can be absorbed, greatly interferes with digestion. Every constipated person should make it a regular practice to take a drink of water about four hours after meals, at bed time, and a half hour before breakfast. Astringent foods such as blackberries, raspberries, blueberries, elderberries, persimmons, quinces, some varieties (puckery ones) of pears are constipating and should be avoided. Those who take but little exercise, should partake sparingly of white bread, eggs, milk, lean meat, sugar, and all alcoholic liquors. These have too little waste material for an exclusive diet, except for such labor as gives great abdominal, as well as general exercise. Torpid bowels must have some stimulation, from bran, fiber, seeds or acids. On account of the fine bran, digestibility, and nourishing properties, the health foods heretofore described are great aids in the cure and prevention of constipation. It is not desirable to use irritating substances to such an extent as to cause the intestines to lose their sensibility. This is very important and often overlooked. Bran is nature's specific for constipation, but as modern man has become accustomed to have all food ground, it is almost impossible to get any one to sufficiently masticate it without regrinding. Where foods containing fine bran cannot be had, it may be washed, boiled, roasted and reground, and then eaten with any food or flavor most palatable. There is no reason why the bran so treated should not be mixed with flour—one part to three or four of flour—and made into bread. Nearly all garden vegetables contain a large amount of fiber, but because of their coarseness they are hard to digest and often fail to produce the desired effect, because of the flatulence they may produce. An intestine distended with gas, cannot contract and propel its contents.

Some foods affect the nervous system of a few persons, so that they act on the bowels at once. Idiosyncrasies must always be reckoned with, but the following are the most useful of the foods that might be classed as laxative:

Water; wheat preparations containing fine bran; rolled oats; bran boiled, roasted and reground; corn bread, corn mush and rye bread; string beans, (fiber); greens, (fiber); strawberries, (acid and seeds); figs, (seeds); apples, (acid and water); peaches, (acid and water); lemons and oranges, (acids); cream and nut oils.

Exercise is a great aid to good health, but its effect on constipation is somewhat exaggerated. There are many whose employment requires great activity, but yet they suffer from constipation. The exercise that directly combats constipation must be abdominal, which requires movements of the body rather than the legs. Striking, bending backward and forward, and kneading the abdomen from right to left and downward, is almost certain to be effective. Alternate hot and cold douches of the abdomen are sometimes very effective. The influence of the mind is far greater than many suppose. A determination to stool at a certain time every day is very important. The modern closet is very faulty as the sitting posture is unnatural. The weight of the body should be borne on the feet and arms without any pressure on the buttocks.

Whatever aids the general health is of value, and must not be lost sight of.

Proper clothing, baths, ventilation of shops and houses, with ample time for sleep, combined with a suitable diet as herein outlined, will make the old feel young, and give the young renewed energy.

Diarrhoea.

This is also a very common complaint, and a common effect of constipation. When fecal matter is long retained in the bowels it may cause an irritation and diarrhoea—nature's way of relieving herself. This is a frequent source of diarrhoea—the kind that alternates with constipation. Other causes of diarrhoea are indigestion, an excessively acid chyme, poisonous ptomanies in decayed food, nervousness, and disease of other organs.

Acrid and solid substances that resist the disintegrating action of the digestive secretions, often cause diarrhoea. The effect of green apples is well known. Green corn and peas often produce similar results. This is because of the

tough cellulose covering that envelops them. Oatmeal
causes a few persons to have diarrhoea and large quantities of sweet cider is almost certain to produce it. What
causes it in one, may have little or no effect on another,
because the intestinal membrane of different persons are
not equally sensitive to irritating substances. In case of
indigestion, mal-fermentation may cause the intestinal
contents to become sufficiently irritating to cause diarrhoea, The diarrhoea resulting from excessive acid chyme
usually appears in the night in connection with an attack
of gastritis, because the stomach secretes too much acid
which irritates the intestines as soon as it passes out of
the stomach. In the Summer season, decaying fruit is
a prolific source of diarrhoea. The toxic substances such
fruit contains, deranges digestion and irritates the mucous
membranes. Diarrhoea from nervous causes is not general, although it is said to be a common thing among
soldiers before going into battle. Persons suffering from
nervous diseases, may be subject to it, but the ordinary
affairs of life do not produce sufficient fright, shock or
nervous tension to cause diarrhoea.

With the exception of nervous diarrhoea, there is always offending matter in the intestines that causes it, and
nature undertakes to wash it out. The use of opiates
and astringents to keep poisonous matter in the system, by
checking the diarrhoea is a striking example of the misuse
of drugs. People who live properly, will not have diarrhoea, but reason should teach them that offending matter
in the bowels must be gotten rid of in some way, and the
cause of the original irritation stopped. If the entire intestinal tract could be irrigated, relief would be quick, but
as this cannot be done, the next best thing is to wash out
the lower bowel and disinfect the upper ones through the
stomach. The lives of many children might be saved in

this way. Where the discharge is so exhausting as to drain the blood, and cause danger of collapse, inject hypodermically luke-warm water to which a little salt has been added. The diet in diarrhoea must be very bland. One of the favorite foods is the flour ball. Take wheat flour and tie in a fine linen or cotton cloth, and then put it in boiling water. Boil eight to ten hours continuously. Flour so treated may then be served with boiled milk. Some physicians advise that the solid part of the flour ball be peeled off and the remainder again boiled for ten hours. It is certain that the second boiling will do no harm, as flour so treated cannot be cooked too much. Next to the flour ball or in connection with it, meat juice may be used. Broil a thick piece of steak lightly and then express the juice.

Albumen water is also very useful. This consists of the white of raw eggs dissolved in water—the white of one or two eggs to a glass of water is sufficient. A little salt may be added.

Beaten or lightly boiled eggs and milk will meet the needs of the system in diarrhoea. All coarse vegetables, sour fruits, sugar, mushes and salt meats must be left out of the dietary. If constipation be the cause of the diarrhoea, the diet must be adapted to it as soon as the acute symptoms of the diarrhoea have subsided.

Chronic Diarrhoea.

This is a disease in which there is chronic inflammation of the intestines. The intestines must be relieved of all the work possible, and digestion performed in the stomach. The diet should be of meat juice and milk, diluted with lime water; eggs and flour ball, for variety. Raspberry and blackberry juice without the seeds may often be employed with good results. Careful attention must be given to general hygiene.

This is a serious inflammation, usually of the large intestines. There is sloughing of the glandular membranes and bloody mucous discharges. Diet similar to diarrhoea.

Piles.

This is an inflammation of the lower part of the rectum, caused by constipation and over-eating of highly seasoned or fatty food, which produces great fullness of the portal circulation. It is also caused by diseases of the liver, heart, uterus, and diseases of other organs. Avoid tea, strong coffee, pickles, sour fruits, all alcoholic drinks, green and canned corn and coarse vegetables. When acute stage is relieved, diet similar to that of constipation will probably give relief.

Injections of hot water are very beneficial, although cold water will sometimes give better results. Avoid straining at stools, keep quiet, and especially keep the bowels active.

Intestinal Indigestion.

This is intended to cover various disorders of the intestines, pancreas and liver, that are mainly functional. The stomach specialist meets a good many cases where the stomach is practically inactive, but intestinal digestion good. The faddists take such examples and construct one inflexible rule, and when applied to other conditions the wonderful cure advocated utterly fails. Another class have good, active stomachs and gastric digestion, but poor intestinal digestion. Such persons have often been treated by various physicians, for many years with poor results, because the stomach was not the seat of the disease. There is such close sympathy between the stomach and the intestines, that it is not easy to determine the real nature of many disorders. If the intestines fail to perform their functions properly, the food may be carried out before digested or may be so long retained as to cause putrefaction. It is obvious that the stomach cannot empty itself when the outlet is clogged. On the other hand, if the stomach fails to perform its work and discharges its contents in bad condition, the symptoms will be very similar to those when the intestines are solely at fault.

The causes of intestinal dyspepsia are very much the same as those of the stomach. In most cases it will be found that the intestines are permanently dilated, and have, in a measure, lost their power to propel their contents. In such cases constipation and diarrhoea will alternate for a time, with a tendency toward one or the other. Among women the excessive use of sweets and strong tea are dietetic factors, but it is probable that corset constriction and the constipation of child-bearing stand out above all other causes.

Among men the use of tobacco and alcoholic liquors, especially beer, are prominent causes.

There will be a feeling of great fullness and weight in the abdomen, three or four hours after meals, accompanied with more or less pain, and there may be occasional vomiting without much nausea. The vomit will usually contain bile enough to give it a greenish cast, and when it is brought up without nausea or retching, it is one of the most characteristic signs of intestintal disorders. Nervous vomiting occurs without great nausea, but it has no time relation to meals, and rarely brings up bile. Violent retching in catarrh of the stomach, or attacks of malaria or other acute infectious diseases, often bring up bile, but is accompanied with intense nausea, and can hardly be confounded with chronic intestinal disorders.

All foods that are hard to dissolve and quick to ferment, must be left out of the dietary, starch and fats are to be used in small quantities, without sugar, syrup, cake, pudding or preserves. All astringent fruits, vinegar, sweet potatoes, mashed potatoes, green or canned corn are like-

wise prohibited.   Cereal mushes and gruels are advocated
by some for all disorders, but they will not do for intestinal dyspepsia.   The cereals must be eaten in the form of
dry unleavened bread, crackers or dry meal twice cooked.

If there is a tendency to chronic diarrhoea, with inflammation and mucous stools, the cereals must be free
from bran, but in most cases fine bran in the bread will
produce healthy action.   When the intestinal starch digestion is bad, it is urgent that only dry bread be used, so that
a very large amount of saliva will be secreted.   This will
insure good salivary digestion, which is always important,
and absolutely indispensable in intestinal dyspepsia.   To
put it in another way: The food must be prepared for absorption in the mouth and the stomach.   Now as the
stomach only digests proteid foods, it follows that they
must be used in preference to starches for a large part of
the diet, and that the starches must be aseptic, extra well
cooked, roasted and eaten dry.   The foods best adapted
are very fine wheat bran, starch (cereals) dextrinized by a
high degree of heat, wheat gluten, malted nuts, cream, nut
oils, fresh lean meat, stewed, roasted or boiled, baked apples, baked bananas, stewed peaches, small quantities of
baked potatoes, fish, milk, eggs, gelatine, bean and pea
soup, when strained.   If the patient is weak, meat juice
will be most serviceable.   Boil a thick piece of steak and
grind and express the juice by as high a pressure as obtainable.   Sweet or very sour fruit must be sparingly
used, when there is pain or great tenderness.   Four to six
ounces of milk and hot water, or milk and cereal coffee
may be taken with the meals, but no other drink.

Diseases of Pancreas.

The pancreas performs the largest part in digestion, but
its abnormalities are the least understood of any organ of
the body.   Owing to its position in the body, it cannot be

examined externally, and internally only after death. Apparently it does not produce any violent smyptoms, but experiments on animals prove that it is indispensable to life, and many autopsies show that the pancreas was the principal organ diseased. It is now believed that diabetes is mainly a disease of the pancreas and liver. Some of the most recent medical writers learnedly describe the various structural changes produced by different diseases of the pancreas, but leaving out conjecture there are only two methods of diagnosing diseases of the pancreas, and that is by chemical analysis of the urine and of the stools. Sugar in the urine and excessive fat in the stools indicate disease of the pancreas. There is not much clinical experience reported in the dietetic treatment of pancreatic diseases, but as both starches and fats are mainly dependent upon pancreatic secretions to render them capable of absorption, it follows that foods of this class must be artificially treated. Fats must be emulsified and as cream is the only natural emulsion it would seem to be a suitable food. Nut or cod-liver oil may be emulsified as follows: Pour the oil in a cup and add about half as much water. Take revolving egg beater, put blades in the oil and water, and then operate as rapidly as possible until the water finely divides the oil, and then add beaten egg without stopping the beating until the mixture is complete.

Starches may be partly predigested by heat (long cooking at high temperature), by pancreatic preparations of commerce, and by malt extracts. (See diseases of stomach.) These methods may be used with great benefit in tuberculosis, anemia, and other diseases, and should be tried whenever there is a suspicion of pancreatic insufficiency. Milk, wheat gluten, powdered meat, eggs and fish should form a large part of the diet.

The liver is the largest organ of the body, and probably performs the greatest number of different functions. It is situated on the right side and extends below the lower ribs, overlapping the pyloric end of the stomach. During the past few years hundreds of experiments have been made to determine its various functions, and from these experiments we learn that the liver secretes from fifteen to twenty-one ounces of bile in twenty-four hours; that bile is composed of mucus, taurocholic and glycocholic acids, bilirubin, biliverdin, cholesterin and the salts of potassium, calcium, magnesium and iron; that bile emulsifies fats and increases the absorptive power of intestinal membranes; that it is laxative and to a limited extent acts as an intestinal antiseptic. The liver changes starch and fat into glycogen, sometimes called animal starch. It also changes albumen into serum-albumen. These are the final changes food receives before being converted into heat and tissue. It is supposed that the liver acts as a sort of reservoir for nutriment, and that it gives out glycogen as the system needs it. Another function of the liver, probably equally important, is that of arresting poisons. When poisons were injected into the veins leading directly to the liver they produced but little or no effect; but when injected in blood vessels leading from the liver, death quickly resulted. The liver arrests poisonous substances generated in the system, destroys the dead tissue of the body, and might aptly be called the supply, and the discharge-center of the blood. An organ with so many functions, must almost necessarily be subject to a great many disorders, so that the common question, "How is your liver?" may be considered almost equivalent to "How is your health?"

Causes of Diseases of the Liver.

The causes are very numerous; but for our purpose, four classifications will be sufficient:

1.   Engorgement from excess of rich food and irritating condiments.

2.   Poisons introduced into the body.

3.   Contagious and infectious diseases.

4.   Poisons generated within the system.

The French are very fond of fat goose livers, from which they make a dish called "pate de foies gras." To obtain the livers they confine the geese separately in small coops and feed them all the fattening foods they can be made to swallow. In a short time the livers of the fowls become three or four times the natural size, and then the goose is killed. A good many people treat themselves as the French do the geese to make fat livers, only they do not kill themselves quite so suddenly as the hatchet does the goose. An excess of fat, such as butter, gravy, fat meats, shortening, with great quantities of other foods, produce great engorgement of the portal circulation, and ultimately partial or total disability of the liver. Man is supposed to be a creature of reason, yet it is difficult to understand why so many poison themselves. It can only be accounted for on the theory mentioned in the beginning of the book, that the animal nature is stronger than the intellectual. The most common illustration of this is found in alcoholic poisoning. The human system can only burn up a small quantity of alcohol, and when a considerable amount is ingested, most of it must be excreted. Just what function the liver performs in this work has not been satisfactorily determined, but when alcohol can be found nowhere else in the system after its ingestion, it may be found in abundance in the liver. It is probable that the liver holds a large amount of alcohol until it can be gradually excreted, and because of this function of the liver it is usually the most injured of any organ in the body when alcohol is used in excess. It is well known that large users of alcoholic liquors quickly succumb to acute infectious diseases, or at least are much more seriously affected by them. This may, in part, be due to congestion of the mucous membranes which alcohol causes, but it is probable that the liver of alcoholics loses its poison-destroying power, and because of this the systems of such persons become infected, and the natural power of resistance greatly lessened. Workers in lead and copper smelters are also much subject to liver diseases. Most of the minerals are slowly excreted, and when taken into the system, either by the mouth or absorbed from handling, there is a gradual accumulation, until the liver becomes almost wholly clogged, which results in disease. We must not overlook the most universal of all poisons—tobacco. Every tobacco user is dependent on the fidelity of his liver to save him from tobacco poisoning, and the injury will be in proportion to its capacity to protect the system. In the treatment of contagious and infectious diseases, the use of tobacco will likely delay recovery. This is also true of diseases of nutrition. The specific infectious and contagious diseases seriously complicate the liver, probably because of the increased destruction of tissue in the body, and the poisonous bacteria and other organisms. The most common diseases of this character are malaria, typhoid fever, scarlatina, diphtheria and syphilis. The fourth cause of liver diseases is the source of most of its ailments, because the changes the food must undergo for the production of heat and the repair of tissue, make possible the constant production of poisonous compounds. The products of every form of indigestion, as well as the non-elimination of effete tissue, are in some degree poisonous. If the liver and other organs be sufficiently capable, the body will be protected from ill effects; otherwise, there is auto-intoxication—self-poisoning.

These are named according to the structural changes
the liver undergoes, the most common of which is jaundice. This is a stoppage in the natural flow of bile and
its absorption into the system, causing a yellowish tint of
the skin. Hyperaemia is congestion of the portal circulation—a blood engorgement. Suppurative hepatitis, is
abcess of the liver, and cirrhosis is a chronic inflammatory
liver disease, characterized by a nodular roughness of the
surface.

Symptoms.

Enlargement and feeling of fulness in the right side.
An unusual fullness of abdominal veins, an irregular and
intermittent pulse, digestive disturbance, loss of appetite,
and especially loathing of fatty foods. Dropsy, jaundice,
difficult breathing, tension in the region of the stomach
and liver, slight chills, sharp pain in the right side, radiat-
ing to right shoulder and a great increase or decrease in
amount of urea.

Diet.

We have seen persons quickly recover from jaundice by
eating large quantities of fresh peaches, after medical treatment had failed to give relief. There is probably no disease where the large use of laxative fruits, such as apples,
peaches, strawberries and oranges have such beneficial effects. The fruit should be fresh and used without sugar.
Sour fruits are never indicated in acute inflammatory condition of intestines, and if complicated with diseases of
the liver, fruits must be kept out of the diet until acute
symptoms subside. A free use of fruits is recommended
by some in alcoholism, and it is claimed, with some reason, that when fruits are plentiful and cheap, the general
use of alcoholic liquors greatly decreases. It is to be
hoped that further observation and experience will prove
this claim and give reliable facts that will be of great value.

In diseases of the liver, as in other diseases, it should be kept in mind that rest for the diseased organ, nourishment and freedom from irritation, are most essential. No food so nearly meets these requirements as milk. Where the liver is enlarged from the excess of rich food, it would seem rational to conclude that a light diet, mostly liquid, should be prescribed; and it is probable that the "fruit cure" for rheumatism and other diseases, rests on this theory. When the liver is affected from such diseases as typhoid or malaria, it is necessary to furnish all the nourishing food that can be digested and assimilated, such as meat juice, eggs, gelatine, gluten and bread. When there is no intestinal inflammation, fine wheat bran that has been roasted and re-ground, will help keep the intestines active.

CHAPTER XXX.
CHRONIC DISEASES.

KIDNEYS.

Acute nephritis is an acute inflammation of the kidneys, characterized by albumen, and other pathological elements, in the urine. The disease may develop into chronic nephritis. Chronic nephritis, or Bright's disease, may also originate without having been preceded by any acute symptoms.

Causes.

Most of the acute cases are caused by the poisonous matter resulting from such infectious diseases as scarlet fever, diphtheria, typhoid fever, small-pox and malaria. Other infectious diseases or diseases resulting from self-poisoning are also causes. Any condition which hinders the functional activity of the skin, or that throws additional work on the kidneys, may cause acute nephritis. This may also be the case in diseases of the skin, extensive burns, or from cold and exposure which arrest its functional activity. The kidneys perform a very important function in removing poisons or poisonous substances from the body, and are, therefore, likely to be greatly affected by them. This is especially true of lead and alcohol in all its forms. Painters and others who work in lead are much subject to this complaint. A distinguished medical writer classes Bright's disease as of uric acid origin, which is but another name for poison resulting from effete matter in the system. In addition to these general causes, Bright's disease undoubtedly has a nervous origin, which comes from shock, emotional excitment and high nervous tension. Hereditary tendencies are very pronounced in some families, as evidenced by the fact that entire families have died of the disease.

Symptoms.

The symptoms are mainly constitutional. In acute nephritis there may be dropsical effusions. These are often manifested by swelling of the face, feet, legs and other parts of the body, with diminished amount of urine. In the acute attacks, there may be chilliness, pain in the back and limbs, dull headache and general physical weakness. In chronic nephritis, or Bright's disease, there may be attacks of nervousness, indigestion, headache, and particularly shortness of breath after climbing a stairway. There is no pain to warn the patient, and it not unfrequently happens that people have Bright's disease and are hopelessly incurable before they ascertain the fact. A chemical analysis of the urine is easily made, and no physician should be permitted to practice medicine who does not make frequent examinations of this kind. In disease, the exact condition of the kidneys is best determined by a microscopical examination.

Diet.

There is no food that gives as good results in diseases of the kidneys as milk. In some cases, it will be best to use it almost exclusively for some months. If it does not agree with the patient, or is not sufficiently nourishing to sustain the strength, well-baked bread, or rice that has been boiled several hours, or until the grains disintegrate, may be used to thicken the milk. Should there be constipation, rolled oats may be substituted for the bread and rice. If the patient loses strength he may be given wheat gluten and powdered nuts and fish. The latter is the only meat ordinarily allowable, but some permit a small amount of chicken. Where the digestive organs are in good condition and the patient has considerable vigor, a moderate amount of green vegetables may be allowed, but never in acute attacks. Neutral fruits, or those somewhat sweet, will not likely do any harm. They should always be baked or stewed. In Bright's disease all irritating substances, such as pepper, mustard and condiments generally, are prohibited. Likewise all meat, except fish, and chicken occasionally; all pungent vegetables, such as raw onions, together with alcoholic liquors of every description, vinegar, sour fruits, tea and coffee. It is of greatest importance that the skin be kept in the most active condition possible. Daily baths, followed by thorough rubbing of the skin for a half hour, is essential to good condition and a prolonged existence. Chronic Bright's disease is classed among the incurable ones, but those who are so afflicted must not worry about it, as it doubles their speed towards the end. With good care, one may live many years with Bright's disease, and be much alive long after some of those who were supposed to be more fortunate, have passed away.

Diabetes Mellitus.

The origin of diabetes has not been satisfactorily determined. It was formerly supposed to be a disease of the kidneys, but modern research has established the fact that the kidneys are merely the intermediaries for excreting the sugar contained in the urine. It is now supposed that the pancreas, liver, and base of the brain, are the principal organs originally affected, and the probable source of this disease. The disease of diabetes is principally characterized by an excessive elimination of urine of high specific gravity, containing a large amount of grape sugar—technically called glycosuria.

It is dependent upon a morbid condition of the system, which prevents the grape sugar contained in the foods being properly taken up by the different organs of the body. Starch is converted into grape sugar, which is greatly increased in diabetes by the constant use of starchy or saccharine foods. This is due to perversion of storage capacity of glycogen in the liver and muscles, and insufficient consumption of sugar in the tissues, because the cells of diabetic patients are unable to perform their sugar-consuming functions. The causes are heredity, infectious diseases, diseases of the pancreas, great mental worry, over-eating and shock to the nervous system. The disease is most common in Southern Italy and India. Of the races, the Jews are most subject to the disease.

Diabetes is fatal in children and is exceedingly dangerous to those so afflicted under the age of thirty, and they especially should completely exclude from their diet-list all starches and sugars. In those over forty-five it is not so dangerous to health as to allow them the moderate use of starchy foods. It should be borne in mind, that it is more important to keep up the strength of the patient, than to reduce the sugar in the urine.

All diabetics should be unconditionally allowed the use of all varieties of meat, such as beef, veal, mutton, fowl, game, pork, tongue, brain, sweet-breads, kidneys, marrow bones, cured meats, fresh fish, shell-fish, preserved fish, oils, eggs, milk, cheese, nuts (except chestnuts), lettuce, endives, spinach, onions, leeks, asparagus, cabbage, meat soups without sugar or flour. Fats and oils are especially useful in this disease.

Gluten and bran bread, without the starch of the flour, may be used, but ordinary bread sparingly, if at all.

Sugar, jellies, sweetmeats, pastry, sweet wines, all flour, cereal or other starches, potatoes, honey, sweet fruits, are forbidden.

Diabetics must be kept free from worry or exhausting labor, and should drink carbonated waters, lemon juice, weak coffee without milk or sugar, but saccharin may be used. Whiskey is frequently useful, but no beer. Regular habits are of greatest importance.

Tuberculosis.

Consumption is one of the greatest enemies of the human race. Its slow insidious attack, has for centuries kept its infectious nature in the background, but thanks to modern research, with the aids of the microscope and medical science, its real nature is much better understood.

It is best known by the masses as a disease of gradual emaciation of the body, suppuration and wasting of the lungs, accompanied by cough. The disease is caused by a microbe called bacillus tuberculosis. Those who possess a high degree of physical vigor, seem to have greater immunity from this disease, than those who are weak. The bacillus may be transmitted to children, but they are not likely to live long. It is quite probable that hereditary weakness causes a pre-disposition to this disease, but the fatalities of consumptive families are due to contagion rather than hereditary tendencies. One consumptive in a family, furnishes infection for all the relatives, and sooner or later, some one will be sufficiently weakened to furnish a lodging place, and ultimately become a victim of the tubercle bacilli. This is the principal reason why several or all of some families die of this disease.

While consumption is caused by the tubercle bacilli, indirect causes are insufficient nourishment, bad ventilation, hot gas-lit shops, ulcers in the throat, and moist atmosphere.

The disease is said to sometimes arise from the milk of tuberculous milk cows. Bronchitis is said to cause twelve per cent of the cases, but the cause of bronchitis is also due to bad hygienic living. The tubercle bacillus measures about one eight thousandth of an inch in diameter, and two or three times as long as thick. The disease usually begins at the apex of the lungs.

The early symptoms of this disease are indigestion, failure of appetite, repugnance to fats, exhaustion on slight exercise, slight fever, night sweats and expectoration. The formation of a cavity is generally followed by regular morning expectoration, and after this night sweats, slightly elevated temperature in the afternoon, loss of flesh, weight and color, the drawn look of the face, the hectic spot on the cheek.

In first stage there is constipation, third stage likely diarrhoea.

Diet.

As tuberculosis is a disease dependent on sub-nutrition, its cure is pre-eminently dependent upon forced feeding. The tubercle bacillus will not stay in good blood for a great length of time. As a rule, the patient loses appetite, eats but little fat, and as a result the tissue of the body is burned for heat. Now when the system must use part of itself to furnish heat, it can be readily understood how consumptives grow gradually weaker and less and less able to throw off disease. The diet must therefore be rich, and ready for assimilation, and nothing meets this demand as well as whisky and emulsified fats, which may be either milk and cream, nut oils, or cod-liver oil. The patient should take all the milk possible. Some of the methods heretofore described will insure success. Some prescribe raw beef steak, but meat powder is far better. It should be made of chicken or beef, always from fresh meat. Meat broths, and a diet of "slops" will not do. As soon as the stomach will digest rich food, powdered nuts may be added to the dietary, but so long as the system is weak, nut butter is much more easily assimilated.

Next to meat powder, beaten eggs, is the best proteid

food. Constipation can be avoided by using the entire
grain of the cereals, wheat, oats, rice and corn. They
should be boiled for some hours dried and roasted, and
then ground to fine flour. If desired, they may again
be cooked for a few minutes and served with milk, eggs,
or meat powder, or flavored to suit.

It will aid digestion to eat a good deal of dry food. If
the cereals prepared as described do not prevent constipation, bran should be treated as heretofore described in
diseases of the stomach, and used with each meal. Coffee
is only permissible for flavor, and tea not at all. Fried
foods, coarse vegetables, raw vegetables, pickles, pastry,
and doughy bread, candy, salt meats, cheese, and condiments should form no part of the dietary. Should there
be sour stomach, or flatulence, sugar must be omitted,
otherwise it can be used in a moderate way. Fruit juices,
peaches and cooked apples may be used freely in most
cases.

The patient must take all the food that can be used in the
system, but never gormandize. If it is too much trouble
to take so much pains with the diet, don't do it, but order
your funeral outfit. Don't be foolish enough to rely on
drugs. They are useful adjuncts, but good blood alone,
can cure.

Insomnia.

Sleeplessness is most usually caused by some disturbance of circulation—often the result of nervous excitement. Anything that produces great activity of the mind,
such as grief, joy, business cares, anger, stimulants, or
drugs, may produce insomnia. It may also be caused by
great fatigue and pain, but the principal cause is indigestion. Disorders of nutrition may both be the cause, and
the result of insomnia.

The diet, of course, will depend much on the cause. If

it be from an excess of food-engorgement,—give nature a chance to unload. If the cause be of a social character, get rid of the cause, and by all means leave off tea, coffee, tobacco, and liquor, or at least, reduce the quantity to a nominal amount. In cases of long standing, where there is physical weakness, massage, in connection with proper diet, will work wonders. In ordinary cases, a brisk walk will draw the blood from the head and regulate the circulation. When there is apparently good health, and no great amount of nervousness, a light supper at 6 P. M. with a glass of milk—malted milk preferable—or a baked apple, or even a piece of bread before retiring will be sufficient to insure sleep. If the food supply be regulated according to the needs of the system as already explained, the worst cases will readily yield. (See Dilatation of Stomach.)

Diseases of the Heart.

The heart is the engine of the human body and its attachments are pipes with valves quite similar to those of ordinary pumps. The valves prevent the return of the blood as it is forced forward. The principal heart troubles are:

(1)   Obstruction or displacement of the heart.

(2)   Enlargement of the heart and changes in its structure.

(3)   Leakage of the valves of the heart.

(4)   Failure from over-stimulation, exertion, or from lack of healthy blood.

(5)   Disease from excessive use of tobacco or alcohol.

The blood vessels cause disease and death from:

(1)   Rupture.

(2)   Obstruction to flow of blood.

(3)   Increased resistance within the vessels.

Probably the most common of all heart ailments result

from displacement of the heart due to excessive flatulence of the stomach. If the stomach becomes inflated like a balloon, it pushes the diaphragm upward against the heart. This causes the greatest anxiety and distress until relieved, and is supposed to cause many deaths, where the patient was ordinarily well and ate an extra hearty supper, but was found dead in the morning. Displacement or obstruction from tumors or water surrounding the heart, must also have a very serious effect.

All these produce sensations of fainting, difficult breathing, and a feeling as though death was imminent.

Enlargement and changes in structure cover a large field in heart troubles. It includes general enlargement, thickening of the walls, increase or decrease in size of cavities, aneurisms, changes due to cancerous growths and inflammations of the heart and connecting membranes.

Anything that interferes with the circulation of the blood may cause heart enlargement. If the heart be stimulated to twice its usual work, or the blood vessels obstructed so that more force is required to make the blood flow through them, the heart will increase in size.

Enlarged hearts are found in those who drink quantities of alcoholic liquors, over-exert themselves (as bicycle riders), and those who have contracted blood vessels due to poisonous matter in the blood.

Strong pulse, easily flushed face, headache, dizziness, shortness of breath, disturbance of digestion.

Fatty Degeneration.—In this disease the structure of the muscular walls is changed by part of the fibers disappearing and in their place, globules of fat are deposited.

Causes.—Beer drinking, excessive use of alcoholic liquors and ice water, excessive corpulency and anaemia are also causes.

Symptoms.—Short breathing, flabby tissue, nervousness, irritability of temper, dizziness and frequent fainting.

Palpitation of the heart, and fainting may arise from displacement, caused by a distended stomach.

In valvular diseases of the heart, the changed structure of the valves may prevent their fully opening or closing, so that it takes more power to force the blood through the arteries, and when there is leakage, it is more than double work, for a large part flows back. Any one who has operated a leaky pump will get something of an idea of this condition. The symptoms are much like other diseases of the heart—attacks of fainting, giddiness, shortness of breath, and the necessity of keeping the head elevated when lying down, and in advanced cases, swelling of the feet, face, poor circulation, weak pulse and dropsy.

A large number of deaths result from heart failure. This may happen because of excessive stimulation, so that when the stimulant is withheld the fatigue of the heart results in its stopping.

The same effect may result from poisonous effects of drugs and from long-continued fever. The higher the temperature the more frequent the beats of the heart. Probably more deaths result after fever has left the patient than before. This is usually called exhaustion, but means that there was no nourishment for the overworked heart, Any weakness of the heart due to disease may cause heart failure.

There is still another form of heart trouble known as nervous palpitation, common to narrow-chested and nervous people, especially those addicted to the use of tea and coffee. There is no organic disease in this class of cases, and all that is necessary is to eat a plain diet with care as to exercise, and avoid excitement and anything that prevents sleep.

The blood vessels are liable to disease somewhat similar to the heart. In the disease called aneurism, the walls of the vessels become stretched in places, so that rupture is liable to occur at any time.

Obstruction to the flow of blood may come from tumors, from pressure resulting from displacement of organs in the body, and from accumulated matter in the bowels, which may cause piles, or other diseases. Obstructions may also occur from tight clothing, hats, corsets, waistbands, and shoes.

The first and most essential thing is to remove the producing cause. Quit liquor drinking—and all other bad habits. If it is from "scorching," quit it. If from contracted blood vessels, eat little or no meat, and wear loose clothes.

In heart disease it is necessary to reduce the volume of blood and increase its quality. This will require a dry diet, and a small quantity of fluids. This is especially true in obesity. The diet may be similar to that in catarrh or dilatation of the stomach, with the exception that the quantity of liquids be much less than in either of these diseases. Where it is complicated with obesity or diseases of the kidneys, the diet must meet the conditions in these diseases.

Rickets.

Rickets is a disease essentially due to improper food, but influenced by unsanitary surroundings, such as filth, bad air and water. It is due mainly to deficiency in mineral matter and develops in children who are fed on sugar, condensed milk, sterilized milk, fat, and starch foods, that contain but little or no mineral salts. The disease also occurs in children occasionally, whom their mothers nurse —due to some deficiency in the mother's milk. Children who are permitted to eat fried foods, pickles, beer, green and over-ripe fruits, and indigestible foods generally, are subject to rickets.

There is frequently vomiting in the earlier symptoms of rickets, which indicates digestive disturbances.

Rickety children are listless and peevish when awake, and restless when asleep. The bones become soft, and if the child walks, it becomes deformed, twisted or bow-legged, and the spine may become curved. There may be emaciation, or the child may be fat or flabby. See diet for children.

Anaemia and Chlorosis.

This disease has attracted a great deal of attention and besides the medical literature on the subject, short articles have frequently appeared in the newspapers and magazines. These have usually been misleading, so that it is looked upon as a disease of the blood, rather than impoverished blood, which it really is. Anaemia is not a curse sent from Mars or Jupiter, but the natural result of plain, every-day ignorance; or at least indiscretion in diet and habits. If the system is not supplied with the necessary elements to make good blood, or the blood be poisoned by effete matter in the system, or drained by profuse discharges, anaemia results. It is most common in girls during puberty; also frequently found among young women—especially students—and a little less frequent among women generally.

Causes.

The chief causes are: insufficient clothing on arms and legs, too little exercise, lack of pure air, and, above all, a diet in which candy, pickles and pastry form the larger part. Secondary causes are profuse discharges (which are also due to errors in living), absorption of pus from suppurating inflammations, drugs, and possibly from eating too little food of any kind. Women may have anaemia, sick-headache, bilious attacks, female complaints, or other disorders, and persist in saying that nothing they eat "hurts them," which may be literally true, but not true in effect. There is something remarkable about the perversion of young girls' appetites at puberty, because the more anaemic they are, the more they crave injurious substances. Parents should bear in mind that poorly-nourished girls will be imperfectly developed women, who, in turn, will probably become mothers of degenerate children. Many make a great mistake in supposing that fat is an indication of good blood and vigor. The test of good blood is health, strength and energy. Many anaemic children are unjustly called lazy, while in fact they have no vital force. They merely exist in form, but not in an active one. The discussion of foods and dietaries in part one, thoroughly covers the subject; but attention cannot too often be drawn to some errors, and among them is the habit of girls "piecing" between meals, eating fried foods, pickles, pastry and white bread. A plain, well-cooked, cereal diet, with stewed or roasted meat, milk, cream, soft-boiled or poached eggs, ground nuts, without strong tea or coffee, will soon dispel anaemia. The time is coming when it will be odious to be sick.

Epilepsy, or Fits.

Epileptic fits have many causes. When due to pressure on the brain from injury to the skull, the remedy is only a surgical one. The chief cause, however, is probably due to uric acid in the blood, and the fits become a habit of the nervous system. In this class of cases it is merely another manifestation of the same thing that produces sick-headache, asthma, rheumatism, and kindred diseases, although epilepsy is not nearly so common. When one or both parents are troubled with sick-headache, asthma, or rheumatism, and a child has epilepsy, it raises a strong presumption that it is of uric acid origin. In all diseases of this class, but little or no meat should be eaten, and care taken not to eat an excess of starch or sugar. There will be more or less indigestion, which must be treated according to the conditions found. Daily baths and exercise in the open air will be very beneficial. If the patient be weak, the baths should be tepid until cold ones can be borne. Constipation must be avoided, by using fine cereal bran. A glass of water should always be drunk an hour before meals, and at bed time. A vegetable diet, fresh air, an active skin and bowels, and alkaline waters, will do much for epileptics.

Asthma.

Asthma is the spasmodic contraction of the breathing tubes, which prevents the free entrance and exit of air. The attacks come on more or less irregularly, and may be brought on by a number of causes, such as strong odors, dust, bad air, and by inflammation in other parts of the body. Asthma belongs to the arthritic diseases, and is caused by some defect in the elimination of waste, particularly effete tissue or excess of tissue-forming foods. Another common term for tendencies to diseases of this class, is uric acid diathesis, which is believed to be caused by imperfect excretion of uric acid. Whenever there is an excess of uric acid in the system of those who are predisposed to asthma, it contracts the blood vessels of the air passages. Just how the results are brought about is more or less a matter of conjecture, but it is probable that when the blood is laden as described, that anything which slightly affects the nerves of the bronchial system, will bring on an attack of asthma. Where the uric acid diathesis exists in a family, one may have asthma, some rheumatism or gout, some sick-headache, some other diseases of the same class, such as some form of epilepsy, eczema, dyspepsia, throat diseases, etc. Medical treatment of asthma is only palliative. The only substantial benefit asthmatics can receive is through their diet, and place of living. This is not always an easy matter to regulate, as there is frequently a dilated stomach, and most asthmatics are obese. Leaving off tea, coffee, ale, wine, beer, meat and sugar, will greatly benefit and probably cure those who have good digestion. If the patient be thin and have poor circulation, it may be necessary to prescribe whisky, but no other liquor; especially none of a fermented character. The diet should mostly be restricted to wheat, oats, corn and rice, prepared as in diseases of the stomach. Thoroughly cooked cereals, eaten dry and well masticated, furnish the best diet. If more fat is needed, use cream and powdered nuts, but never without grinding as fine as flour. Milk is also permissible, but where the stomach is dilated it will need to be modified in some of the ways heretofore explained. Fruits are usually prohibited, except for flavoring, although neutral fruits, such as sub-acid apples and grapes, may be eaten during good health. Water, milk, sassafras tea and cereal coffee are the only drinks permissible to use.

Leanness.

The doctor's advice to the fat and to the lean, has long been a target for the humorous paragrapher. It is just possible, too, that they draw a picture too often true, when they describe the doctor as advising the fat patient to leave off starch, sugar and fat, and the lean one to eat them. Leanness cannot be cured by any rule of arithmetic, but only by scientific dieting. People may be lean because they eat too much fat and starch, as well as not enough. It is a matter of digestion rather than ingestion. Leanness is undoubtedly hereditary; but Nature never intended one to be too lean for vigor and endurance. Capacity for work and general health is the real standard for condition. When people fall below their average weight, with a tendency toward weakness, there is cause for apprehension.

Causes of Leanness.

Besides hereditary tendencies, mental worry, over-exertion, mental or physical, loss of sleep, inability to digest starchy food, insufficient, or too much food. Those who are too thin, or lack strength, but are otherwise well, should reckon just how much food they consume each day, and if the quantity eaten does not produce at least 3,000 calories of heat (see dietaries) for moderate work and average size, the diet is deficient. Food in great excess causes indigestion, which may prevent the formation of fat. Such persons will likely have sour stomachs and heartburn, with gaseous eructation (see gastritis and dilated stomach). Those who have excessive acid secretions will not have a sour stomach from fermentation, until the stomach becomes dilated. Persons of this tendency are nearly always hungry, and are sometimes charged with "eating so much that it makes them poor to carry it."

Diet.

The first requisite is freedom from worry or mental strain. Then regular habits and plenty of sleep. Ten hours' sleep is a great aid toward the accumulation of fat. There must be no excesses of any character, and two or three moderately cold baths (in a warm room), should be taken every week. After each bath crash towels or flesh brushes must be used for at least ten minutes, until the skin glows. People who are "run down," should not usually be put on large quantities of starch and fat. The system must be toned up by moderate quantities of food that are easily digested. Malted wheat gluten and beaten eggs, with well-cooked wheat foods, containing fine bran, will secure activity of the bowels and put the system in condition. Cream, nut butter, and malted nuts will fatten the quickest of all foods. It is a common notion that both milk and water are fattening. The ingestion of large quantities of water may cause more fat to be stored in the system, but it could not, of itself, make fat; and milk is not ordinarily more than three or four per cent. fat. Starch digestion will greatly be increased by using dry food. Tea and coffee should be dropped in favor of hot water and milk, or cereal coffee. The quantity should not exceed four or five ounces at a meal. Particular care should be taken to dress warmly. If the leanness be due to diarrhoea or female diseases, or, in fact, any disease, they must be treated accordingly. Tobacco users should quit the habit, or at least use the least possible.

Obesity—Corpulence.

Obesity is the accumulation of an excessive amount of fat in the body.

Causes.

Its most usual cause is over-eating, although some obese people eat very little. In most cases there is a hereditary tendency to corpulency, which readily develops when the diet and habits favor it. The most fattening foods ordinarily used are fat meat, butter, lard, or other fat used in cooking, cream, sugar, bread, potatoes, the cereals and nuts. The yolks of eggs should also be included. Water does not produce fat, but favors its accumulation. Alcoholic liquors, especially beer, produce some fat, and besides being fattening, they cause tissue changes and the deposit of fat that would otherwise be burned up. Muscular inactivity aids in the accumulation of fat, because fat is consumed by muscular exercise. Those who are anaemic often become fat because poor blood will not carry enough oxygen to burn up the elements that make fat.

Those who are fat and anaemic suffer intensely from ex-posure to cold.

Effects of Fat.

An excess of fat affects the system in the following ways:

(1) It prevents the radiation of heat; (2) interferes with the action of the muscles and various organs of the body; (3) increases the volume of blood; (4) obstructs the circulation; (5) changes the structure of the heart and liver and weakens their action.

The first symptom that plainly indicates injury from an excess of fat is an increased rate in breathing from slight exertion, and later without any exertion at all. This con-dition is due, (1) to the fact that the heart cannot force the blood through the lungs fast enough: (2) to the re-stricted action of the lungs. The accumulation of fat in the abdomen prevents the descent of the diaphragm and the full expansion and contraction of the lungs.

An excess of fat is a common cause of heart failure and apoplexy. The increased volume of blood and the in-creased resistance to the flow of blood overwork the heart. This is noticeable when an obese person rapidly climbs a hill, or even a stairway. There will be a throbbing of the heart, a fullness of the head, and a fainting sensation.

Dietetic and Hygienic Treatment.

Many cures for obesity have, from time to time, been advocated, but almost all of them at the expense of diges-tion. A good many women resort to vinegar drinking, without much reduction of fat and probably great injury to their digestive organs. The use of cathartics is objec-tionable for the same reason, so that the treatment for this disease mainly comes to a restriction as to food and drink, and sufficient exercise to burn up the excess of fat. The ordinary foods that produce fat, are starch, all the cereals, sugar, syrup or sweetened foods, cream, butter, fat meat, lard and nuts. Whether meat from which all fat has been removed would produce fat has not been satisfactorily determined, but it is generally believed that it will not. Single articles of food at each meal have often been recommended. Only one good effect could possibly result from this, and that is, that the appetite would be quickly satisfied and only a small amount of food eaten. Such a dietary may cause disease because there is no certainty that the necessary food elements would be supplied. Obesity is often difficult to treat, because obese persons frequently have idiosyncrasies, and the disease is seldom found without complications. The diseases obesity seems to favor are gout, rheumatism, asthma, heart diseases and dyspepsia. Rheumatism and gout require plenty of water and a vegetable diet. In such cases, the diet should consist mainly of such garden vegetables as string beans, beets, cabbage, cauliflower, celery, stewed onions, lettuce, spinach, turnips, parsnips, and carrots. All should be well cooked and chopped crosswise of their fibre. For the tissue-forming foods, fresh water fish, skimmed milk, the whites of eggs, and prepared wheat gluten. Two or three ounces of entire wheat bread, or potatoes may be allowed each day. If this diet does not make the bowels active use plenty of bran and wheat middlings, which should be boiled, roasted and re-ground as fine as possible. It may then be made into cakes, but no shortening should be used. If obesity is not complicated with gout, rheumatism, or asthma, lean beef, mutton, veal and chicken may be added to the dietary and milk, except for flavoring, taken from it. Water unites with other substances to form fat, and except where there is some disease such as rheumatism, that requires a large amount of water to carry away effete matter, the dryer the diet, the more rapid the reduction in weight. The object is to consume more water than is taken into the system, thus compelling the use of water already in the body and the burning up of accumulated fat. All fried foods are prohibited, because of the fat used in cooking. One ounce of butter a day may be allowed if no cream or shortened foods are eaten. Three or four ounces of weak coffee, water, milk and water, or cereal coffee, at each meal is all the fluid that should be drunk at meals. A small quantity of water between meals is allowable. It is necessary to eat some starch and fat, and to take fluids, but the quantity consumed must be much below an ordinary diet. Gluten biscuit, made by the Sanitarium Health Food Company, should be substituted for bread, if circumstances will permit.

Mountain climbing, gymnastics and Turkish baths are advocated for obesity; but, before any vigorous exercise is undertaken, it would be well to ascertain how much the heart will stand. When there is no danger of heart failure, plenty of bodily exercise, with restricted diet, will quickly reduce fat. The fat-reducing value of Turkish baths is greatly over-rated, because the water loss from the sweating process is likely to be soon replaced. The baths are useful to remove effete matter and aid in maintaining a dry diet without injury.

Headache.

This ailment has so many causes that a complete description of them would fill a volume; but they may be briefly described by saying: that headaches are caused by poisonous substances in the blood, and by some disturbance in circulation and diseases of the nervous system.

The blood may contain toxic substances from indigestion, effete matter from incomplete elimination, or from the various micro-organisms that produce contagious or infectious diseases.

The periodical attacks of sick-headache are usually due to excess of uric acid in the system. The only cure known is to live mainly on a cereal diet, take out-door exercise, plenty of water and daily baths.

Disturbance in circulation results from disease, mental excitement and pressure from clothing. Headaches so produced can only be cured by removing the causes that produce them.

See diseases of the stomach, intestines, liver, asthma, rheumatism and epilepsy.

CHAPTER XXXI.

ACUTE DISEASES.

Cold.

Cold is an elastic term that is applied to a large number of symptoms, varying much in severity. The most common form is called coryza, but better known as "cold in the head." This form of cold is an acute inflammation of the mucous membranes of the nose and adjacent passages. The swollen membranes cause an oppressive sense of fullness in the head, and may close the air passage in one or both nostrils, which makes it necessary to breathe through the mouth. At the beginning of the attack there will be a watery discharge from the nose. As the more acute symptoms subside, the discharge becomes thicker, and sometimes quite hard.

Tosilitis—Quinsy.

This is another manifestation of cold, but instead of the inflammation of the membranes of the nose, it is an inflammation of the throat and tonsils. Children are much more subject to the disease than adults.

Pharyngitis—Sore Throat.

Pharyngitis is an inflammation of the membranes of the throat, and is a common form of cold.

Acute Bronchitis.

This is an inflammation of the lining membranes of the trachea and bronchial tubes—the air passages of the lungs. It may follow a cold in the head, sore throat, or the cold may first affect the bronchial membranes. There will usually be a feeling of constriction in the front of the chest, difficult breathing, and a pronounced cough, although the cough may be a symptom in other diseases, especially from the throat. All of the membranous inflammations incident to cold, may become chronic, if the causes producing them are constant, or even frequent.

Causes of Colds.

Colds are caused by chilling the surface of the body, especially after being overheated. Cold, damp atmosphere, insufficient clothing, chilling the skin, overheated and badly-ventilated houses, are all causes of colds. It is likely that over-eating and constipation are more frequently the cause of colds, than is supposed. Whatever disturbs the circulation of the blood and prevents the elimination of waste may cause a cold.

Hygienic Treatment.

Colds should be prevented by proper living, but when once contracted, how shall we get rid of them? The answer is very simple: remove the cause by restoring the functions of the skin, and other excretory outlets. This can best be done by vigorous exercise sufficient to start profuse perspiration. Turkish, vapor, or other baths, that open the pores of the skin and cause free perspiration, will cure a cold at the beginning of the attack, and shorten one already existing.

After a sweat, the skin should be cleansed, and sponged at least three times. The first time with tepid, then cool, and finally with moderately cold water. This must be followed by thorough rubbing of the skin, dry clothing and a temperature moderately warm for several hours, or patient may go to bed and keep warm. The bowels must be kept active and houses well ventilated. Cold packs with dry covering, give great relief from cough and discomfort in the face.

Diet.

In an acute attack it will be well to eat but little. The maxim "feed a cold and starve a fever" would be better if rendered: "If you will feed a cold you will have a fever to starve." The diet in ordinary acute cases should be laxative (see constipation) and reduced one-half for two or three days. In chronic cases, where the patient is weak, a rich diet should be allowed and the patient fed on well-cooked cereals, gluten, eggs, milk, powdered meat and powdered nuts.

Malaria.

It is now generally accepted as a fact, that malaria is a germ disease. Where there is rich land and heavy vegetation, there will likely be malaria about the end of the Summer, and in hot climates all the year. There is also more or less malaria adjacent to streams, and it is believed that it always exists in newly-cultivated land.

Symptoms.

Languor, headache, aching of body and limbs, chilly sensations, followed by fever. There are many types of malaria manifested as "dumb" ague, daily, alternate, and third day ague. Also many forms of intermittent fever. It is supposed that the germs of different types of malaria require different lengths of time for development. At a certain stage, they produce the acute attacks with chill, high fever, perspiration. When the fever subsides, the symptoms may disappear until more germs are matured.

Diet.

It is remarkable that so little attention has been paid to the dietetic treatment of this disease. Good blood and an active liver resist malaria without any drugs, but this fact seems to have been lost sight of in its treatment. We have seen patients treated for months with constant recurring attacks, without any notice being taken of the fact that the patient was living on fried pork, hard-fried eggs, hot biscuits, fried potatoes, and strong coffee. No one can eat such a diet and keep well, much less get well, when debili-

tated by malaria, which engorges the liver, impoverishes
the blood and weakens the whole digestive system. After
an attack of malaria, the system is a much damaged fort-
ress. The blood is the agency of repair, and food the material. The stomach and bowels will need to be cleansed
and disinfected, and as soon as the fever is down, easily
digested, and non-fermentable foods should be given, such
as egg punch, beaten egg, in three parts milk and one part
cream, that have been sterilized, or pasteurized. Gelatine
may be used instead of egg, where more agreeable. These
may be flavored to suit. In most cases sour fruit, such
as oranges, lemons, peaches, baked apples, strawberries,
and fresh grape juice will give good results. The diet in
convalescence should be similar to that in catarrh of the
stomach. All coarse, tough, or indigestible substances
and fermented foods must be avoided. The cereals should
be well cooked and malted. Baths and general care will
greatly aid. Get all foul and effete matter out, and good
healthy blood as soon as possible, and malaria will seek
weaker victims.

Scarlet Fever.

Scarlet fever is a contagious and infectious disease, and
is an inflammation of both skin and mucous membranes
of the body. It has three periods:

1st. Invasion, which lasts from 24 to 48 hours.
2nd. Eruption, which lasts from 5 to 7 days.
3rd. Desquamation, from the 7th to the 21st day.

Eruption commences second or third day after fever,
and consists of very numerous points about the size of pin
heads. Between these the skin is of natural color. As the
eruption develops, the red points unite, but fade in from
five to eight days.

Symptoms.

Pain in the back and limbs, coldness of skin, headache,

nausea and vomiting, followed by sensation of heat and high temperature, often accompanied by delirium. In severe cases, the tongue is swollen and presents a strawberry appearance. Symptoms increase in severity as eruption appears. The urine is scant and of dark red hue. The nervous system and kidneys are most affected by the scarlet fever poison. The disease can be communicated by personal contact, by atmosphere, clothing, animals, or food, especially milk. The scales are the most contagious. The darker the color of the eruption the more severe the disease. Measles, or erythema, are liable to be mistaken for scarlet fever. There is this difference: In scarlet fever the eruption first appears on the neck and chest, while in measles, first on face. Eruption does not always appear, and in such cases it is difficult to distinguish it from diphtheria. The urine of scarlatinous patients should be carefully examined every day after the eruption has appeared, as it not infrequently happens the kidneys are badly inflamed, and if not watched may result in Bright's disease and death.

Diet.

This disease is so frequently a source of kidney disease, that great care should be exercised in feeding, until recovery is complete. Milk is the best food. It may be diluted with well-cooked gruels, but not with gelatine or other animal food. In serious cases milk should be the principal food for some weeks. Effervescing waters, barley water, orange and fruit juices (except astringent ones —raspberries, etc.), may be given to moisten the mouth and quench the thirst. During high fever the patient will take from two to five ounces of fluid every hour. In using animal foods during convalescence, eggs, fish and chicken should be allowed before other meats.

Diphtheria is a specific infectious disease caused by a microbe known as Klebs-Loeffler bacillus. It is locally manifested by an intense inflammation of the throat, with constitutional symptoms, due to poison produced by the bacillus. Infection may occur by being near the patient, or may be carried by healthy persons to others. Many cases occur by relaxing rules of precaution after patients seem to be about well. The virus attaches itself to clothing, bedding and the room in which the patient has lived.

Symptoms.

The period of incubation is from two to seven days. There is slight chilliness, aching pains in the body and limbs, followed by fever. Temperature usually rises to 103, and in severe cases 104, the first twenty-four hours.

In addition to the danger to life which the diphtheretic throat may cause, the kidneys are liable to be seriously affected, so that the diseases which result indirectly from the poisoned condition of the blood need to be carefully guarded against. The urine should be examined daily for kidney complications.

Diet.

Diphtheria is the most malignant of the common diseases, and needs especial care in feeding. Plain ice cream without sugar is both nourishing and soothing to the throat. Repugnance to food is a bad diagnostic sign, and every effort possible must be made to overcome it, by offering a variety of flavors. Foods thickened with cream, beaten eggs, or gruels, will sometimes be more easily swallowed than either milk or water. If there is a feeble pulse and danger of heart failure, alcoholic stimulation may be required. In such cases, egg-nog and milk punch should be given.

Haemoptysis—Haemorrhage of the Lungs.

Haemorrhage of the lungs or blood-spitting, has many causes:

1. Rupture from external violence, as from blows or falls.

2. Violent exertion, as an attempt to perform some extraordinary feat, and inflammation from any cause, throwing an excess of blood to the lungs.

3. Secondary effect of heart disease, pressure of tumors, or enlarged glands.

4. The perforation of blood vessels by disease.

In hemorrhage of the lungs the blood is coughed up, not vomited. In either case it may be possible for the blood to come from the throat. The patient must lie flat on the back without pillow, and must not move or speak. Food must be administered with a spoon. As the volume of blood must be kept as small as possible, but little fluid should be given. Use cracked ice to quench the thirst; alcohol may do harm. Beaten egg and meat powder, with small quantities of milk, should form the principal part of the diet. All fluids must be given cold, or only lukewarm. The prepared foods may be given with milk. Should there be nausea, rectal feeding must be substituted to prevent retching or vomiting.

Measles.

Measles is a contagious and infectious disease, manifesting itself by an eruption of red spots accompanied by catarrh of the air passages and more or less fever. The eruption makes its appearance first on the face, then upon the neck, chest, over the body, and lastly upon the back of the hand, which usually requires four days from first appearance on the face. As it disappears it assumes more of a yellowish red. The spots are crescent-shaped, and from one-eighth to two-fifths of an inch in diameter, and are usually bright red. Sometimes the eruption is so thick as to entirely cover portions of the skin. In severe cases, where there is hemorrhage, black measles develop. Average period of incubation is eight days.

Symptoms.

The first symptoms, eight to ten days after exposure, is a languid, chilly feeling, and in young children, convulsions occasionally occur. There will be pain in front part of the head and general feeling akin to a severe cold in the head, and likely a constant watery and irritating discharge from the nose, with sneezing and coughing. Fever will be developed the second day and continue two to four days.

Diet.

Similar to that used in fevers.

Pneumonia.

This disease is an acute inflammation of the general structure of the lungs, which may invade any part of the entire lungs.

Causes.

It is probably due to a germ, but as it cannot find lodgment in a healthy person, it may be said to be due to exposure, cold and wet, bad air, over-eating, impoverished blood, especially where it is an incident to malaria, and neglect of skin. Habits which allow waste to accumulate in the system, make pneumonia possible with but very little exposure. It is now conceded to be an infectious disease.

Symptoms.

It is usually preceded by a cold with accompanying aches and pains; these are followed by a chill, and a rapid rise of fever.

Diet.

Pneumonia is a disease of short duration in acute form, but it needs careful dietetic treatment. Vomiting must be guarded against. Milk, meat juice, the white of egg beaten, and whisky are mainly relied on. Cereals cooked as heretofore described may be malted and given in form of gruels, without sugar. In convalescence an easily digested and nourishing diet will be necessary.

Skin Diseases.

Skin diseases are caused by parasites (such as itch), contagious and infectious diseases, diseases of the heart and blood vessels, nervous disorders, but most commonly by some form of starch or fat indigestion, or deficient elimination of nitrogenous waste. The principal investigator of uric acid diseases, classes skin eruptions among them.

Erythema, or Urticaria (Hives).

This is the most common of all skin eruptions. There is also a form known as nettle rash, so well known it needs no description.

Causes.

They are caused by some article of diet, most usually oysters, lobsters, strawberries, bananas, sausage, rich gravy, mushrooms, cheese, and sometimes sour fruit. Bathe the eruption with soda water—small teaspoonful of soda to pint of water—and eat a plain cereal diet.

Acne.

This is an eruption of red pimples on the face, that do not readily disappear.

Causes.

Excess of fats or starch, doughnuts, sausage, fried meat, buckwheat cakes, or griddle cakes, pastry, excess of sugar. All over-rich or indigestible foods are bad and should be left out of the diet; also tea, coffee and alcohol in all forms. Little liquid should be drunk at meal time, but a glass of hot water a half hour before, when practicable, especially before breakfast, and before retiring at night will be beneficial, if not long continued.

Eczema.

This is the most common of all skin diseases not of a transient nature, and begins with an inflamed patch which

often spreads.   There are usually red pimples, but the red
spots may only be swollen vesicles with watery discharges
followed by thickening, scabbing, scaling and intense itching.

Cause.

Excessive meat eating and other causes enumerated in
acne.   The same rules apply to diet.

Yellow Fever.

Yellow fever is an infectious disease of a violent character that is caused by a specific germ which thrives in
animal and vegetable matter.   It is essentially a disease of the tropics, and is rarely observed above 40 degrees
north and 20 degrees south latitude, and is always checked
by cold weather.   It is usually spread from one part to another by ships.   The period of incubation is from twelve
hours to four days.

Symptoms.

Commences with a chill, alternating with flushes of
heat, gradually settling down into a regular fever.   The
skin varies from dark or swarthy yellow, to dark orange;
bowels usually constipated in the beginning, followed by
violent diarrhoea.   As the disease progresses, pain in the
stomach and bowels become severe, and they are sensitive
to pressure.   The most pronounced symptom is the black
vomit, due to hemorrhages from the violent inflammation
of the stomach, intestines, kidneys, spleen and liver.
Nothing but predigested food should be given, until the
most severe symptoms have subsided.   Then the diet must
be soft and easily digested.

Laryngismus Stridulus—Spasmodic Croup.

Spasmodic croup is the ordinary croup, in which the
spasm affects the muscles of the larynx and makes breathing difficult, causing a wheezing sound at each respiration.
The disease seldom affects any but children, although

hysterical persons and grown-up people having a catarrhal inflammation of the mucous membrane, are sometimes subject to it.   It is said to be an ailment of the nerves, and is entirely reflex, so that the real trouble is to be found elsewhere, and most likely deranged digestion.

Causes.

Over-feeding, improper food, constipation, colds and teething. (See infant feeding and dietaries for children.)

Septicaemia (Blood Poisoning).

This is a constitutional disease due to poisoning from the absorption of pus into the blood.   Bacteria are always present and enter the system from some local injury or decomposing tissue in the system, such as typhoid ulcers, sloughing membrane of throat in diphtheria, abdominal abscesses, decomposing placenta remaining in the womb after child-birth or miscarriage, suppuration in small-pox, and especially wounds made in handling dead bodies which are in an advanced state of putrefaction; also likely to result where a large part of the skin has been burned, and from inflammations where pus is formed in considerable amount.

Symptoms.

Decided chill and rise in temperature, but often irregular chills, followed by profuse and exhausting night sweats. Skin soon becomes dry and hot; pulse 120 to 140, small and intermittent. Tongue at first coated with a white fur, later becomes glazed, dry, grayish-brown and cracked; skin slightly jaundiced, and usually diarrhoea.

Prevention.

Wash wounds with water that has been boiled, and in any disease which pus is formed, care must be taken to have it removed, that it may not be reabsorbed.

Diet.

Malted milk, malted cereals, pancreatinized meat powder, or eggs, and beef blood.

Whooping cough is an acute contagious disease and is primarily a catarrhal bronchitis or specific catarrh of the mucous membranes of respiratory tract, and attended by a peculiar laryngeal and bronchial spasm. It depends on a specific germ given off by the breath and conveyed through the air to the healthy. Incubation varies from five days to two weeks. It may be carried by clothing, and contracted by breathing infected air. The fever in early stage is intermittent, but great languor and restlessness are common.

The diet should be wholesome. (See infant feeding.)

Scrofula.

Scrofula is a disease which manifests itself in various parts of the body and is doubtless a blood disease. It most usually breaks out on the skin, but may affect the mucous membranes, bones, tissues, glands, and in fact almost any part of the body may become diseased from scrofula. The tubercle bacilli are found in scrofulous sores, and it is not known whether it is the cause, or whether it appears after the disease, but it is now believed that tuberculosis and hereditary syphilis cover most or all cases of scrofula. Both the disease and a scrofulous tendency seem to be hereditary, as well as acquired, from improper feeding, and unsanitary surroundings. Scrofulous children usually have white skins, delicate blue veins, large, lustrous eyes and show an irritable, nervous disposition and premature brightness.

Diet should consist of cereals, milk, eggs, and nuts.

Pleurisy.

Pleurisy is an inflammation of the investing membrane of the lungs. It is caused by some functional derangement of other organs of the body.

Symptoms.

Pleurisy usually begins with some sharp, stitch-like

pain in the chest, which, for a time, increases with each breath. The pulse becomes quickened, the breathing is rapid and difficult. Temperature ranges from 100 to 104 and is usually accompanied by a short, dry cough, that is very distressing. The attacks sometimes begin with a chill, and such cases are difficult to distinguish from pneumonia.

Diet.

The derangement of other organs that cause pleurisy should be ascertained and the diet adapted.

Erysipelas.

Erysipelas is a contagious and infectious disease caused by micro-organisms. It usually first appears in wounds, but not always; for parts of the body supposed to be healthy may be first attacked. The local manifestations may be found in any of the lining membranes, but it is more likely to affect the skin and tissues beneath. It is highly contagious among surgical cases and women in child-birth. Buildings may remain infected an indefinite period. It may also be transmitted by atmosphere, clothing, and in other ways. The part of the body having local symptoms will have a deep rose color, and it may be distinguished from rheumatism by the rapidity with which the inflammation spreads.

Diet should be light and nourishing; such as milk, cream, meat juice, beaten or lightly boiled eggs, cereal gruels and nut puree.

Apoplexy.

Apoplexy is a haemorrhage or stoppage in the blood vessels of the brain. It may be preceded by dizziness, and sense of discomfort, or it may come suddenly. There is loss of consciousness and frequently death in a few minutes depending on the size of the haemorrhage or clot on the brain. Paralysis is likely to follow, although it sometimes comes from softening of the brain and from diseases of the spinal cord. Great care is required in feeding, and food must be given in teaspoonful doses, and the ability of the patient to swallow noticed. In some cases it may be necessary to put a tube down the throat and feed through it, or feeding by the rectum may be resorted to. Give milk or egg lemonade, or beaten eggs and milk.

Mumps.

Mumps is a catarrhal inflammation of the parotid glands and may affect either or both. It is generally regarded as contagious, and is first manifested by swelling of the gland beneath the ear, fever and stiffness of the jaws. Frequently there are pains in the limbs and chilly sensations.

Diet.

Milk, gruel and broths. No solid foods should be given and the starches for gruels must be extra well cooked.

Tetanus—Lockjaw.

Lockjaw is a disease of the nervous system due to some specific bacillus that enters through a wound. It has received its name doubtless because of the muscular spasms which first affect the muscles of the jaw, and prevent opening of the mouth. It is supposed to be a disease of the spinal cord, but the change in its structure is so slight that examinations made after death do not fully reveal its real nature. It may result from as slight an injury as a splinter in the hand or foot, or from a small cut; may also attack women who have had a miscarriage or been confined, and newly-born children. The spasms have been known to be so violent as to break bones, and in bad cases patients have been drawn into an arch, bearing all their weight on the back of the head and heels. Such patients must be fed through a tube inserted between the teeth or through the nostrils. (See fevers.)

Prevent Lockjaw.

It is important that all wounds be cleansed, especially those caused by anything which may be poisonous, such as rusty nails or splinters from wood that has come in contact with dirt. The wound may be washed with water which has been boiled, then cleansed with turpentine—one part turpentine to ten of water.

Typhoid Fever.

Typhoid fever is a continuous fever, caused by an infectious poison, supposed to be due to a micro-organism, known as typhoid bacillus. The fever usually lasts about a month.

How Acquired.

It is believed that drinking impure water is the most usual source of typhoid fever, although epidemics have been traced to food, such as infected oysters and milk. The length of time required for its incubation is not very definite, but is supposed to require two or three weeks to develop; sometimes longer. The first symptoms are languor, slight headache, pain in the limbs, muscular weakness, and a general feeling of indifference and malaise. These sensations are likely to increase with the disease until the fever becomes quite manifest. It must be remembered that many other ailments begin with similar symptoms so that it is difficult to determine with certainty when typhoid fever exists, until the more pronounced symptoms, peculiar to typhoid appear. The most important of these perhaps are (1) temperature. In typhoid it rises with remarkable regularity from day to day, and is from one to two degrees higher in the evening than in the morning. (2) Pale red spots. Generally at the beginning of the second week, a number of small pale red spots called roseolae appear on the skin, especially on the chest and abdomen. There is also sensitiveness in right illiac region. The fever is now well established and all the premonitory symptoms will have disappeared. The face of the patient will be flushed and will likely have a bright patch on the cheeks. Sometimes there is constipation but usually diarrhoea.

Effects of the Disease.

The typhoid bacillus attacks Peyer's patches and solitary follicles in the lower end of the ileum, just above the illeo-coecal valve—the entrance into the large intestines.

About the end of the second week ulcerations are formed, where the bacilli are supposed to multiply and their poison taken up by the blood. The drowsiness and delirium characteristic of typhoid is the result of bacterial poison.

From this brief explanation it will be readily understood why typhoid is such a serious malady. In severe cases the mucous membrane erodes to such an extent, as to ulcerate and cause haemorrhage. This is not necessarily fatal, but is often so.

The ulcers sometimes perforate the intestines and recovery in such cases is very rare, indeed. .

Besides the direct danger from haemorrhage, patients die from exhaustion, and complications of other diseases, especially pneumonia, caused by the poisoned condition of the blood.

In feeding typhoid cases the following facts must be borne in mind:

(1) The patient's strength must be maintained.

(2) The introduction of insoluble food, which may cause perforation, is strictly prohibited.

(3) Food, which, owing to the diseased condition of the patient, cannot be digested, will probably cause fermentation and undue distention of the bowels, and haemorrhage.

The matter of diet in typhoid fever is so important as to deserve extended notice. Most typhoid patients die of exhaustion or perforation of the bowels, which emphasizes the importance of using great care in feeding. Milk may disagree with patients and resort must be had to other foods. Where there is nausea and foul stomach, lavage is often practiced with great benefit.

Cereal gruels, cooked four hours or more and strained, are often well tolerated, especially if malt extract be added (made by steeping commercial malt in cold water for twelve hours). Fruit juices will make the various foods more palatable, and give variety. The beaten white of eggs in water, malted gruels, fruit juices and milk, buttermilk, koumys, about cover typhoid dietaries, except the prepared foods, which are often prescribed.

The lower bowel may be evacuated, if there be constipation, by using an enema of tepid water, to the amount of one to two pints, which should be injected high up in the rectum.

In convalescence, no solid food can be given for at least ten days or two weeks after the fever has ceased. Ignorant but well meaning people have caused the death of many, by suggesting that this or that will not hurt the patient. They do not know that a little solid food or gaseous fermentation may cause perforation of the bowels, and cause the death of the patient. To prevent imprudence in diet, convalescents should be constantly watched, and no food or fruit should be left in their rooms. Pain and distention of the bowels call for immediate medical attention.

Influenza—(La Grippe).

Influenza is an infectious disease characterized by catarrh of the mucous membranes of the air passages, and alimentary canal. It is contagious, and the bacilli are often present in great numbers, even after the severe symptoms have subsided.

The period of incubation is one to four days. Attacks begin with slight fever, chilliness, headache, depression of spirits, pains in various parts of the body, and watery discharges from the nose.

La grippe is liable to affect any organ of the body, and particularly the kidneys and nervous system. Isolation should be practiced when possible, and old people and invalids should be especially guarded from infection. Death may result from heart failure or pneumonia.

Diet.

In this disease it is desirable for the patient to eat as much wheat phosphates as possible, as nearly all persons who are seriously afflicted with la grippe have impoverished blood. If no diarrhoea exists, wheat bran boiled four hours, dried and roasted until brown and then ground to fine flour and eaten with milk, or milk and beaten egg, will give surprising results. As soon as improvement begins the patient should have all the nourishing food that can be digested. Cream, nut oils, malted nuts, cereal foods well cooked, dry bread, toast and meat powder may be added to the dietary. The excessive use of coffee helps keep up the nervousness in this disease.

CHAPTER XXXII.

WHAT TO DO IN ACCIDENTS AND
EMERGENCIES.

When a large blood vessel is opened, death may so quickly result, that it is very important to know what to do to stop the flow of blood, without delay. When there is a cut or wound, the blood may either be a bright scarlet or darker hue. The first is arterial blood and comes from the arteries direct from the heart. The darker is venous, comes from the veins, and is returning to the heart. Various means of arresting the flow of blood may be tried. The most generally useful one and the one most readily applied, is compression. Pressure may be applied in various ways. The pressure of the hands may be used to control the flowing blood until other means more effectual can be secured. Anything in the way of a belt cord, strap, or handkerchief, when drawn tight enough above the wound will stop the flow. Even a rope of hay or grass quickly twisted together will answer, if in the field or on the highway. Should the pressure not be sufficient, it may be increased by placing a short stick under the bandage and twisting the bandage upon itself. Other methods are sometimes necessary and may be tried. The old time remedy of a spider web is probably a successful, but by no means safe method, for while it may arrest hemorrhage it may poison the wound. Turpentine may be used and besides tending to check hemorrhage it also cleanses the wound. In case of punctured wound where pressure cannot be effectually applied, the wound may be plugged with clean linen, saturated with turpentine and water, one part to ten. Strong salt solution, may be used in absence of turpentine. Cold, in the form of ice, or very cold water may be applied to wounds and adjacent structures.

Hemorrhage from the Nose.

Hemorrhage or bleeding from the nose, while rarely a dangerous symptom, is frequently so severe as to cause great anxiety to the patient and to his friends. Slight hemorrhage sometimes occurs in those of robust constitution, and in a few moments ceases spontaneously. Such cases need no treatment, but where the bleeding is frequent, or in great quantity, the nostrils should be examined and the cause, if possible removed. The immediate methods popularly supposed to be efficient is snuffing cold water up the nostrils. This is a measure of doubtful benefit. Pressure on the soft part of the nostrils for about five minutes will frequently control the most severe hemorrhage. Vinegar is sometimes used for the same purpose with good effect. Water may be used at times with excellent results, and does most good in the form of a hot foot bath. Cold to back of neck is also good.

Hemorrhage from the lungs may be confounded with hemorrhage from the stomach, but the following points will serve to distinguish between the two: That from the lungs comes on suddenly or with but little warning to the patient. The blood is coughed up and usually preceded by a tickling sensation in the throat; the blood is bright colored and frothy. Hemorrhage from the stomach is usually preceded by a long existing stomach or intestinal trouble. The blood is usually vomited up and mixed with particles of food, but is not frothy. In pulmonary hemorrhage the patient should be placed at rest, and ice or cloths dipped in ice water, applied to the chest and neck. Stimulants should not be given. The recumbent position should be maintained and the patient forbidden to move a muscle. Life is not often endangered and slight hemorrhages cease spontaneously.

In hemorrhage from the stomach, the patient may be laid across the bed with the feet hanging down, and as in pulmonary hemorrhage, should be kept perfectly quiet. Small pellets of ice may be swallowed and cold cloths placed over the stomach. Iced alum whey should be given every few minutes.

Fractures.

By fracture, we mean a break in a bone or cartilage. When the ends of the bone are driven through the skin, or an opening made that permits the atmosphere to enter, it is compound, and where a nerve, artery or vein is injured, a complicated fracture. The immediate treatment is much the same. The first thing to do is to place the limb in as nearly a natural position as possible. This should be done carefully and gently lest surrounding tissues, arteries, and nerves be injured. After placing the limb in position it should be so maintained by means of splints, or pieces of board. When a leg has been broken, it may be tied to the other. If on the road or in the field, and no other means are obtainable this will be found a very useful expedient. A broken leg requires that the support should extend at least to or above the middle of the thigh. The latter would be better.

When the thigh is broken, if it is possible to secure one, have the splint extended to the armpits. A broken forearm should be bent at the elbow and extended across, and the palm of the hand toward the body, and should be maintained in this position. A fractured arm may be tied to the body, or to a splint extending its whole length. If bandages cannot be secured use may be made of handkerchiefs, or even ropes of straw, or grass may be made to do duty. Where ropes or short bandages are used, they should be tied above and below the fracture, leaving it unbound for two or three inches either way. The splint for the thigh should be tied to the body just below the armpit, around the waist, and several times between the waist and foot. A shirt may be torn up for bandages. In removing the clothing from a broken limb it is best to cut it.

Poisoning.

Poisoning requires prompt and effective treatment, and the patient's life depends largely upon the length of time that has elapsed between the taking of the poison, and the administration of an appropriate antidote. The first and most necessary thing in poisoning by opium, arsenic, phosphorus, or strychnia, is to empty the stomach. The patient should be encouraged to take large draughts of mustard and water, if readily obtainable, but it is better to use salt water, or even warm water, than wait for something better. If none of these are at hand, or do not produce copious vomiting, it should be induced if possible, by tickling the throat with the finger, or better still, if at hand, a feather. Should the patient's hair be long enough, it will probably do as well as a feather. Large quantities of liquids hold the poison in solution, distend the stomach and make the vomiting more effective, if quickly produced. When opium has been taken, and the stomach emptied, the patient should be given frequent draughts of strong coffee, and kept constantly moving about, if able, but if not, apply cold water and maintain artificial respiration. In strychnia poisoning, administer sedatives. The best ones for this purpose are bromide of potassium of sodium, and should be given at once in forty grain doses. If neither of these are quickly obtainable, preparations of opium such as laudanum of paregoric—usually found in every household—may be substituted. In cases of poisoning from arsenic, rat poison or paris green, the patient should be given sulphate of magnesia and salt, and kept at rest. The antidote for phosphorus poisoning, after vomiting, is a small dose of sulphate of copper, and a large dose of magnesia. Do not give oil. Milk and whites of eggs are of value in most cases of poisoning. In poisoning by acids, and alkalies, we have not only the systemic effects to deal with, but also their more painful and destructive local action. They very rapidly destroy the lining membranes of the mouth, oesophagus and stomach, and if their action is not quickly arrested, eat rapidly into deeper tissues.

Vomiting in these cases should not be employed, because the burning of the oesophagus incident to vomiting, will likely do greater harm. Chemical antidotes should be given as quickly as possible. The violent efforts at vomiting may cause rupture or hemorrhage of the stomach, if the destruction of tissue has been extensive. In poisoning by mineral acids, solutions of bicarbonate of soda (baking soda), or chalk, should be given, and when these cannot be obtained quickly, plaster from the walls may be used. In carbolic acid poisoning, give sulphate of magnesia or soda, and olive oil or melted fat—lard, butter, etc. In poisoning by lye give weak acids, such as vinegar or lemon juice, and fats and oils.

Burns.

Burns and scalds are painful accidents of frequent occurrence. They may vary in extent from a slight burn, destroying or inflaming a small portion of the skin to very extensive ones which destroy all the tissues of a part. The first thing to be done is to remove the clothing. This should be done as carefully as possible, so as not to destroy the blister formed. The burned parts should be shielded from the air at once. For this purpose, sprinkle

over the burned surface, until it is completely hidden, flour
or baking soda.   White lead makes an excellent covering.
After a burn has been dressed, it should then be covered
warmly, but lightly.   Care should be taken to prevent
pressure on or near the burned part.   The use of cold
water is beneficial and grateful to burns.   The part or
parts affected, may be immersed in cold water as soon
as possible after the burn, and kept there for from fifteen
minutes to half an hour, depending upon the intensity of
the burn.

Linseed oil and lime water or bicarbonate of soda, may
then be applied to good advantage.   Any pure oil is useful, as it protects the burn from the air.

Burns from acids may be relieved by applying baking
soda or soap; burns from lye may be relieved by vinegar.

The effect of prolonged cold is to stop the circulation of
the blood, which is followed by loss of feeling in the injured parts.

The circulation in the part frozen, should be re-estab-
lished gradually, and this is best done by keeping the
patient in a cold room and rubbing vigorously with snow
or cold water.   In severe cases of freezing, there is danger of gangrene of the part affected.

Hysterics.

Hysterics may be defined as a nervous explosion.   It is
probably best treated in mild attacks, by taking no notice
of it, or by attracting the patient's attention to something
in which she may be much interested.   Rubbing the limbs
and chest will afford great relief.

In fainting, nature assures the patient taking the best
position possible, from the fact that the patient falls and
lies in a horizontal position until consciousness returns.
If for any reason, the patient has not fallen into such a
position, he should at once be laid down,

Do not attempt to raise a person who has fallen in a faint, but let him lie, and loosen the clothing about the neck and chest.

The face may be sprinkled with cold water.

Drowning.

All clothing should be loosened. The patient should be placed over a barrel, or the feet elevated, in fact, stood on his head for a few minutes; this should be done even in the seemingly most hopeless cases, and no effort should be spared in the attempt to restore consciousness. After he has been rolled on the barrel, or otherwise emptied of water, artificial respiration should be tried. The patient should be laid on the back, with shoulders elevated, a coat, shawl, or stick of wood, will answer the purpose. Anything that may have found entrance into the mouth should be removed by inserting the finger, and it would be well to always make an examination in this way. The tongue should then be drawn out of the mouth and held.

To practice artificial respiration, kneel down above the head, grasp both elbows, bring them horizontally from the sides over the head, until they almost meet, then pull strongly for a few seconds, return them to the sides and press with force against the ribs. The movement should then be repeated, until natural respiration has been restored. As soon as possible the patient should be given stimulants. It is always important that he be kept warm.

Gas Poisoning.

Persons are overcome by the impure air of deep wells, and other foul places, as well as by illuminating gas. The treatment is the same in all cases of this character. The patient should be placed in the open air, and if in a room, the windows and doors should be opened at once, the patient's clothing loosened, and artificial respiration as in drowning practiced. Whisky or brandy may also be given.

Epilepsy.

Epilepsy requires the same treatment as fainting, except that care must be taken to prevent biting the tongue; this may be done by placing a handkerchief or piece of wood between the teeth.

Shock.

Shock follows injuries from violence. The patient should be covered with warm, but light coverings, and the bodily temperature maintained by means of hot water applications. Bottles or tins filled with water and placed at the feet and armpits and other portions of the body, is the most convenient form of applying heat. The patient should lie with head lower than the body. Stimulants in the form of liquors or strong coffee may be given, and if the patient is unable to swallow, the liquor may be mixed with water, one part to three, and injected into the rectum.

INDEX.

www.ingramcontent.com/pod-product-compliance
Lightning Source LLC
Chambersburg PA
CBHW032303280326
41932CB00009B/679